Intelligent Networks

Jan Thörner

Artech House
Boston • London

Library of Congress Cataloging-in-Publication Data
Thörner, Jan
Intelligent networks/Jan Thörner
Includes bibliographical references and index.
ISBN 0-89006-706-6
1. Telecommunication systems. 2. Telephone systems. I. Title.
TK5102.5.T536   1994   94-15140
621.382–dc20   CIP

British Library Cataloguing in Publications Data
Thörner, Jan
Intelligent Networks
I. Title
621.38
ISBN 0-89006-706-6

© 1994 ARTECH HOUSE, INC.
685 Canton Street
Norwood, MA 02062

International Standard Book Number: 0-89006-706-6
Library of Congress Catalog Card Number: 94-15140

10 9 8 7 6 5 4

Intelligent Networks

For a complete listing of the *Artech House Telecommunications Library*, turn to the back of this book

To Ulla and our daughter Sandra

Contents

Preface

The worldwide telecommunication network is the greatest and most complex system that exists. Telecommunication is one of the most important, most expansive, and most powerful tools in the 90s, in almost any business area, and it is going to play a much more important role in the private domain. Many services that, today, are provided by other media, in the 90s, will be taken over by telecommunication networks. To fulfill all the requirements this implies, a quite new approach to building, maintaining, changing, and providing services is needed. A much quicker reaction to market demands as well as opportunities to customize services will be other natural consequences. The solution to fulfilling all those requirements is intelligent networks (IN), a concept that was introduced in the 80s and will be widely used in the 90s in all networks.

The first phase of intelligent network implementation is more a new way of thinking or of structuring than a new network. The basic functionality in intelligent networks, the *service control function* (SCF), which hosts the control (or the intelligence) for handling services, will be implemented in *service control points* (SCPs) or in *service switching and control points* (SSCPs). Before intelligent networks, this functionality was spread to several nodes in the network, for example, for a freephone number (i.e., 800 number). So, normally, when IN is introduced into a network, it brings about not a sudden increase in total network intelligence, but rather a new structuring of the already existing one. An increase in intelligence occurs first in the subsequent phases of intelligent network implementation, when such features as mobility, flexibility in charging and billing, advanced customer control facilities, and advanced routing are introduced. These enhancements would, however, not be possible without the new structuring introduced in the first phase.

It is not a coincidence that the intelligent network concept was born in the 80s and not before. It was in the 80s that networks reached, on a widespread scale, the level of technology required for intelligent networks to be introduced, namely, a broad usage of stored program control (SPC) exchanges, digital transmissions, and modern signaling systems like the Common Channel Signaling System (CCSS) No. 7.

The goal of this book is to provide a broad knowledge of intelligent networks, by exploring both the theoretical models of the International Telecommunication Union (ITU-T) (autumn 1993) and the practical experiences of implementing an intelligent network on a real network. Theoretical discussions accompany examinations of the impact of network implementation. User-related aspects of the intelligent network, from the subscribers', the service providers', and the network operators' views, are covered. The book examines the advantages of using intelligent networks as well as the risks to be faced. It also attempts to describe what intelligent networks represent in the long term—a flexible intelligence for any access form and any network.

Chapter 1 defines two major problems facing all networks in the 80s and 90s: (1) updating software in large, complex systems, like a telephone network, and (2) increasing the use of existing services. Chapter 2 takes a more general look at the intelligent network concept, including both practical issues and standardization in ITU-T and ETSI. The chapter also discusses the likely future of intelligent networks. Chapter 3 describes various network implementations of intelligent networks as well as the charging and billing functions. Chapter 4 describes the services available in the first years of intelligent networks, including services outside the intelligent network platform. Chapter 5 explores the risks and threats to intelligent networks that implementers face if they are not careful. Chapter 6 takes a step into the future and considers some new services and functions that will develop as a result of user demands: enhanced mobility, service interworking, and charging and billing. Chapter 7 examines the impact of new services on future IN platforms. Resource allocation and mobility are covered as well. The chapter also discusses the various ways that cooperation between intelligent network platforms can be accomplished. Finally, Chapter 8 speculates on how intelligent networks can help solve the problems raised in Chapter 1.

Acknowledgments

This book would have been much more difficult to write without the support I received from my employer, Telia, and from all my colleagues there. I want to thank Telia for its financial support and thank all my colleagues, who have been involved in one way or another, for their encouragement and support. However, the ideas and opinions expressed in the book are entirely my own and are not meant to reflect the policy, position, or opinion of Telia.

I also want to thank the International Telecommunication Union-Telecommunication Standardization Bureau (ITU-T) for allowing me to use two figures and a table for which they are the copyright holder (Figures 2.1 and 2.2 and Table A.1). The choice of this material is my own and therefore not the responsibility of ITU-T in any way.

Chapter 1

Building, Operating, and Using Large, Complicated Systems

1.1 LARGE SYSTEM EVOLUTION OCCURS IN WAVES

Defining a large system is always difficult. Nevertheless, a large system *can* be described as a system that possesses at least one of these characteristics:

- Requires considerable software to operate it,
- Is extremely complex,
- Covers a large geographic area,
- Contains a large number of users.

Additional criteria for describing large systems may exist, but these are generally the first characteristics that come to mind.

Some large systems are standalone systems, complete with built-in, independent intelligence. These systems may and often do communicate with other systems, but their main characteristic is that they are able to work autonomously, without intervention from the outside world. Excellent examples include airplanes, ships, and space shuttles, all of which are very complex systems that have nearly all control logic or intelligence on board.

Other systems feature distributed intelligence, whereby a single piece of equipment is only a small part of a larger, complex system. The entire system works by regulated cooperation among all the equipment inside, and, more importantly, it often runs in an operation mode that does not allow the whole system to be taken down for a change of hardware, testing, or another reason. The best example, of course, is the worldwide telecommunication network. A telecommunication network must function 24 hours a day, 365 days a year, which means that maintenance and improvements to the system must be carried out during operation.

Looking back at the evolution of large systems, an interesting phenomenon can be observed: large systems tend to evolve in waves (or follow a distinct wave form).

- *A time to sow (spring).* Phase one is the investment in and development of a basic system (a platform). In this phase, no results are visible but considerable amounts of money are normally spent.
- *A time to reap (autumn).* In phase two, investments must be exploited to earn back the money spent during phase one. This is also the phase in which visible results are produced, and the money required to maintain the system is less than the money spent in phase one.
- *A need to sow (spring) again.* There will come a time, however, when we can no longer continue to run the existing system (existing platform) because it is becoming old-fashioned. Consequently, we must open our wallets again to invest either in improving the existing system or in developing a completely new one.

Consequently, the cycle starts all over again. (See Figure 1.1.)

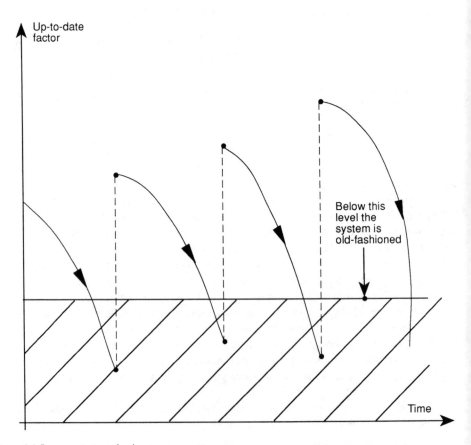

Figure 1.1 Large systems evolve in waves.

The system becomes old-fashioned because general demands increase as the technical evolution continues, and new systems are constructed with greater capabilities. This means that the original system becomes less modern; in other words, the *up-to-date factor* of the system continuously decreases. Once it falls below a certain low level, the system becomes a liability rather than an asset to its owner. Ideally, an organization should upgrade a system or replace it with a new one *before* the system becomes a liability. However, if the system is large and complex, this is a very expensive and time-consuming task. In reality, most systems experience a period, before they are upgraded or replaced, when they are more of a liability than an asset.

Obviously, this discussion is applicable to telecommunication networks. To understand that it can be considered a common trend, we need only look at the U.S. space program:

- During the 50s and 60s, large sums of money were spent building up a *basic system* that would enable human beings to leave the confines of earth.
- The *first visible triumphs*, for the general public, were the first successful launch of an unmanned rocket, which occurred around 1960; followed some years later by the first manned space flight; and the landing of a man on the moon in 1969. In between those events, a large amount of effort was expended, but it was not visible to the general public.
- In the 70s and part of the 80s, the *basic system* was improved.
- The next *visible triumphs* were the ability to reach other planets and the space shuttle, which provided a system that could be used more than once.

Telecommunication networks also evolve in waves, *from the technological point of view*, beginning with *stored program control* (SPC) technology in the 70s, followed by fiber optics and the *integrated service digital network* (ISDN) in the 80s, and continuing with *intelligent networks* and broadband in the 90s. However, this technological progress, where a large amount of money is spent, is not perceived directly by the user.

From the 60s to the 70s, the most visible advancement in telecommunication, *from a subscriber point of view,* was a more reliable network. During the 70s, value-added services, for example, call forwarding and call transfer services, became available in private automatic branch exchanges (PABXs); during the 80s, these services were also available in public networks. Services like freephone (800 numbers) and 900 numbers became available in the public networks in the 80s and 90s (often built first with node-based solutions and not with intelligent network platforms). In the near future, services offered via broadband access will be generally spread among subscribers.

Features visible to the general public are one thing; results that constitute the evolution of a large system are another. Just because results do not become visible at once does not mean that work is not proceeding. In fact, it is often quite the opposite; by the time results become visible, the work is almost completed.

1.2 LARGE SYSTEM DILEMMA

The dilemma of a large system often becomes apparent in attempts to improve its functionality. Users (both providers and customers) of a system have one single overall requirement: to have the best functional system available at any moment.

We must consider three major factors when contemplating improving an existing system:

1. *The cost of improving the system.* Because the complexity of a system increases with each improvement, the cost of making a similar improvement tends to rise more and more each time. (See Figure 1.2.) For example, changing a number plan in one area will become more difficult and more expensive as the number of subscribers in the area increases.
2. *The existing system itself.* The fact that they must improve an existing system instead of building a new system from scratch often poses great problems for project managers. Typically, the existing system was built using old technology and old interfaces, making compatibility a major headache. Consequently, the execution of a project can seldom follow a straightforward approach to the final goal.
3. *The technical evolution in general.* The continuously accelerating speed of technological evolution leads to, among other things,

   - Increasingly shorter intervals between technology generations available on the market,
   - Increasingly shorter time before an operational system becomes outdated.

   The life cycle of a large system can be divided into three phases (see Figure 1.3):

   a) The time it takes a vendor to update an existing system or develop a new one, or the time it takes a customer to search for, evaluate, and install a system. (*Ta*)
   b) The interval during which a system fulfills user requirements. (*Tb*)
   c) The interval during which the system is still in operation but no longer fulfills user requirements. (*Tc*)

If *Tg* represents the *time between the birth of two consecutive technology generations* that are applicable to a given system, the following assumptions concerning evolution and user needs can be made (see Figure 1.4):

1. The faster technology and user needs evolve, the faster we need new systems and the faster a system becomes old-fashioned. This means that both *Tg* (generation

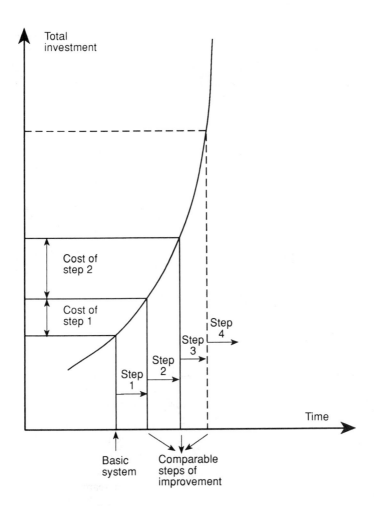

Figure 1.2 The cost of improving an existing system.

time gap) and *Tb* (productive time span) will continuously *decrease* with each new generation of systems.

2. The system becomes more complex, that is, it contains a growing number of new functions with each new generation. At the same time, the number of generations that are simultaneously in operation grows. Both of these conditions result in an increasingly longer development time for the vendor and an increasingly longer evaluation and installation time for the customer for each new generation of systems. Consequently, *Ta* (the development/evaluation and installation time span) will continuously *increase*.

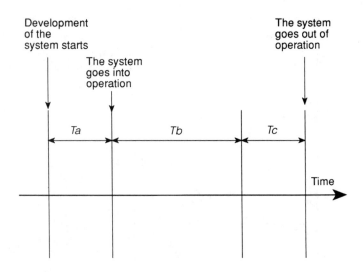

Figure 1.3 The life cycle of a large system.

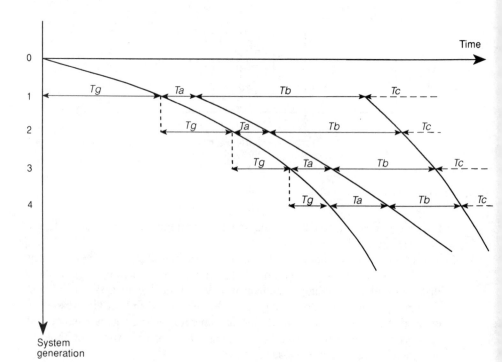

Figure 1.4 The evolution of system generations.

If we look at Figure 1.4, two very interesting questions arise.

1. What if *Ta* reaches the same length as *Tb*, that is, what if the development time or the time required to update a system is as long as the time the system fulfills user requirements?
2. What if *Tg* becomes less than *Ta* (as in system generation 4 in Figure 1.4), that is, what if new generations of systems appear faster than the time required to develop them?

If we do not change the way we work, we will be faced with what I refer to as the *Large SYstem Dilemma*, or *LSYD*. If we do not find a solution to LSYD, we will not be able to continue expanding and evolving systems beyond a certain size, which would be reached very soon.

As can be inferred from these two questions, the key factor is *Ta*, that is, the time required to develop and deploy a new system or to update an existing system to fulfill new requirements. Accordingly, we should be very interested in searching for a way to decrease *Ta*.

Fortunately, such a way exists within the telecommunication area, namely the use of the intelligent network concept. Intelligent networks allow us to decrease one of the most important factors required to upgrade a system to meet user requirements: the time required to introduce new services when demanded. With intelligent networks, the introduction time for a new service is decreased from about two to five years, using conventional software programming and testing, to a number of months. (This is further discussed in Chapter 8.)

1.3 USER WILLINGNESS

The basic point of introducing *value-added services* is to provide extra value for the user and greater revenues to service providers and network operators. These goals are best accomplished by making it easier for subscribers to use their telephone and by enabling different business areas to benefit from using telecommunication services. Put simply, the cost of setting up a telephone call is the same to the network operator, whether the call is successful or not, except that no income is generated in the latter case. The subscriber might also find it frustrating not to reach a called party, getting a busy signal or no reply instead.

If "intelligent" services are introduced to raise the share of completed calls, all parties will profit. That is why all networks give high priority to services resulting in call accomplishment (the calling party really reaches the called party). Examples of services include call forwarding unconditional, remote control of call forwarding, call forwarding if no reply or busy, call completion to busy subscriber, services supporting mobility, and services using time and origin control. Obviously, these services are also highly suitable for most business areas.

But, to make efficient use of a network possessing all these interesting features, subscribers must have a good user interface, which is not the case today. The user interface for the world's largest system, the telephone network, remains *dreadfully primitive*. We still use a star, a pound sign, and the numbers 0–9 (see Figure 1.5). This is, in fact, technology from the 60s in a system for the 90s.

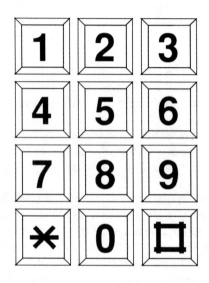

Figure 1.5 The dual-tone multifrequency (DTMF) user interface.

The following hypothetical situation, involving a network operator and a customer, illustrates the antiquated nature of current terminal technology and also suggests how customers can be encouraged to use the network.

- Today the network operator tells a customer, "We have now created a number of services. Please use them!"
- The customer doesn't see all the possibilities offered by the network and replies, "The star and the pound sign are for call forwarding, aren't they? Actually, I never use call forwarding."
- The network operator ponders this and comes up with an idea. "Let's create somewhat more intelligent terminals, where the names of the most common services have their own buttons. This will surely make the terminals more user-friendly, which will increase the use of the services." (Figure 1.6.)
- But, the next time they meet, the customer replies, "It does look good, but I can't see any use for those services in my business."
- The network operator goes home and considers it again. After that, he engages an adviser to help him better understand and respond to the needs of the customer. The results is services that correspond exactly to the needs of the customer.

- The network operator meets the customer. "At last, we have exactly the network services you need for your business." But the answer is not what the network operator expected. The customer replies, "Yes, it certainly seems good, but I already have special routines for this, and you know how difficult it is to change them. It would also cost me a lot of money and time, and my staff wouldn't like to change their working routines right now, as we are in the middle of a very busy period."

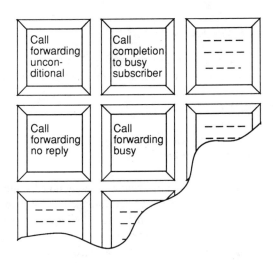

Figure 1.6 A more user-friendly interface.

What is the problem here? We know that

1. The user interface is good, and
2. The services are exactly what the customer needs.

The answer is that the customer is not motivated to change routines and begin using the telecommunication services. Something is missing, something I call: *user willingness* (UW).

Customers require more than good technological solutions to their demands (that is, more than a good user interface and good services), they must be *convinced* by a network operator of the advantages of particular services to be motivated to use them. Furthermore, they need assistance to switch from old routines to new ones, and such help should probably be provided by a network operator.

Naturally, to create UW, the user interface must be a good one and requested services must be 100% fulfilled. But these efforts are obviously not enough. *Exactly how* UW is created is discussed in Section 8.3. For now, we will only observe how very different the situation would be the next time they met if the network operator did succeed in creating UW in the customer. This time the initiative will come from the customer.

- The customer: "You know, call forwarding works fine, but I would also like to have remote control of call forwarding, and I would like it to work in the following manner . . . . When can you provide this?"
- And, upon meeting a colleague, the customer will say, "You know, by using the telecommunication service call forwarding, I have increased sales by \$___ and reduced internal costs per unit sold by \$___."

1.4 THE WORLD'S LARGEST SYSTEM

Section 1.1 identified four criteria of a large system. We can now see that the worldwide telecommunication network possesses all four characteristics. The network

- Requires considerable software,
- Is extremely complex,
- Covers a large geographic area,
- Contains a large number of users.

The worldwide telecommunication network is undoubtedly the world's largest system, and it is no coincidence that considerable efforts have been made to solve the LSYD within telecommunication. As Section 8.2 discusses, the introduction of the intelligent network concept is one important step towards a solution.

Very soon, intelligent networks will offer a much wider range of services, in every network, than was previously possible. The next logical step, therefore, is to focus greater efforts on increasing the use of services introduced by creating both user-friendly interfaces and, of course, UW.

Chapter 2

Intelligent Networks

2.1 NEW TRENDS AND NETWORK ECONOMY ARE THE DRIVING FORCES BEHIND INTELLIGENT NETWORKS

2.1.1 Trends in Society

As *professionals*, we spend a great deal of time establishing and maintaining contacts with other people, no matter what our occupation is. We meet people face to face during meetings, at the office, or in town, or communicate via telephone, telefax, computer mail, and so on.

As *private individuals*, we maintain contacts with relatives, friends, government authorities (local and central), as well as with shops, stores, and other suppliers of what we need to live comfortably.

If we look back 50–100 years, our lives were quite different in many ways. One of the most important basic developments that lifted our standard of living and quality of life was the availability of a fast and reliable telecommunication network, which, compared to other communication media, is significantly less expensive to use.

A person who lived and worked in the United States or Europe 70 years ago maintained some sort of regular contact with an average of, perhaps, 100 people. Today, that number is close to 500. Moreover, 80–90% of those people, 70 years ago, probably lived within 10 miles of each other. Today, 80–90% of our regular contacts live more than 10 miles away, at least as far as business contacts are concerned.

The way we contact people has also changed. Seventy years ago, we usually met face to face. If we used a communication medium at all, it was usually the mail. To make use of the telephone, both communicants had to own one, which was far from certain in those days. Using a phone was also comparatively expensive. Today, the most common way to contact another person is via a medium, such as a telephone, telefax, mail, or electronic mail. The development of a nationwide and, later, an international telecommunication network is one of the factors that, along with railways, roads, ships, and airplanes, spawned the industrial revolution and fueled a rise in the standard of living. All these systems have something in common: they all require substantial investment, and it usually takes considerable time to establish, develop, or change them.

2.1.2 Trends in Technology

A number of inventions and technologies have made it possible for engineers to create fast and reliable telecommunication networks and to continue to increase network intelligence. These include semiconductors (transistors), the ability to make them compact (integrated circuits), digital multiplexing technology, fiber optics, computer technology with increasingly faster and compact *central processor units* (CPUs). Software technology that allows programmers to write programs that, once implemented, can be executed without human control or intervention is another important evolutionary step. These technologies were used to build the SPC exchanges and digital transmission systems that, today, are the foundation of all high-technology telecommunication networks.

Considering these technologies, one word stands out. This word highlights an advancement that has probably contributed the most to enabling us to increase dramatically the standard of telecommunication worldwide. That word is *compactness*. The ability to make technology compact allows us to pack more and more functions close together to speed up internal communication and obtain a better overview of functions. For example, the use of transistors made it possible to reduce the size of a particular function drastically, compared with the old electron tube technology. The integrated circuit gave us a further increase in "functions per space." Multiplexing is another way of achieving compactness by using a single physical connection instead of 24, 30, 384, 480, 6144, 7680, and so on. And fiber optic makes it possible to have a higher bandwidth than copper cables. All these examples were made possible by increasing compactness.

Another important factor is the successive increase in component *reliability*, which, combined with improved design principles, makes it possible for us to create larger and larger systems with maintained good quality and reliability.

2.1.3 Trends in Communication

The way we communicate with each other is also changing. We are abandoning the old network, which, initially, provided only the most basic functionality in terms of making telephone calls. The first changes to that network came as services oriented only to facilitating the placing of phone calls. A service either simplified the task of making a basic call, for example, *abbreviated dialing* and *repetition of the last dialed number*, or it gave assistance in case of unexpected conditions, for example, *call forwarding when busy*. The next step is to introduce services that are not directly connected to a basic call, but rather that give value in themselves. Examples are *televoting* and *premium rate*.

Mobility in a network—first, partial, later, total, and finally, global mobility between geographically separated networks and between networks with different access forms—is a set of evolutionary steps to follow. The access forms used will include not only voice, data, and facsimile, as they do today, but also broadband—in the future, images will be used extensively.

Ultimately, possibilities will spring from the networks. And as they grow, these possibilities will increase demands. The demands will encourage further technical evolution, which will, in turn, increase the possibilities, and so on.

2.1.4 The New Natural Resources

All natural resources have one characteristic in common: they can be exploited to obtain a better living. Often, natural resources cannot be used directly, but must be refined for use. Natural resources are also limited and must be handled with care so we may share them now and in the future. Extravagant use of resources could have disastrous effects on future generations. In the past, natural resources have included iron, wood, water, and air, to mention a few.

In the same way that we must be careful not to overexploit water and air, we must also take care in the way we use telecommunication resources when communicating over distances. These resources are also limited and may not last unless they are handled carefully. Examples of these resources include:

- Radio frequencies
- Telephone numbers (number series)

Two serious problems facing many networks today are that radio frequencies are rapidly filling up and telephone numbers are running out. When a number series in one area is exhausted, adjusting it causes a ripple effect. This means that numbers must be changed for many people, not only in that area but in surrounding areas as well, which is extremely time-consuming and expensive and often causes irritation.

Good *planning* in the use of these resources is essential. One of the most important considerations for network economy in future networks will be *resource allocation and optimization*. (This is further discussed in Section 7.5.) To facilitate current planning of resource use and future resource allocation and optimization, we need an instrument that will provide a good network overview. For both tasks, although they are not its primary aim, the intelligent network can be of assistance.

2.1.5 New Trends and Economic Reasons Require a New Telecommunication Concept

Trends in society, as well as possibilities presented by technological evolution, combined with economic considerations, have fueled development in telecommunication networks, from SPC exchanges in the 70s to the evolution towards the intelligent network in the 80s and 90s.

When we make a telephone call today, we want to either reach a particular *person* or perform a particular *function*, involving, for example, a bank, an insurance company, or a pizzeria. When we make the call, we need to know the exact physical location we are calling so we reach the right telephone. In future, we want to be able to call a person or perform a function by using a *unique number*, no matter where that person or that function is physically located. We will just call the unique number and the network, not we, will know where to route the call.

For some functions, who we reach or which location we call is of no importance, as long as we get the right response and assistance. For example, calls to a bank can be of two types. Either we want to talk to our own personal bank contact (a person), or we simply want to call the bank to discuss general matters (a function). In the latter case, it does not really matter which branch we reach or who answers, as long as we can receive the same assistance from all branches.

On the other hand, if we want to have a pizza delivered, the exact location we call is of extreme importance—unless, of course, we like cold pizza. We need to be sure that the pizzeria that takes our order is located in our general vicinity. Here, the value of intelligent networks becomes apparent. With intelligent networks, we do not need to know where the nearest pizzeria is located. We simply dial the same unique "pizza number," no matter where we are in the country, and the network finds the nearest pizzeria. The result is a hot pizza.

These scenarios, in which a network provides the exact level of service a user desires, illustrate one of the main goals of the introduction of the intelligent network concept. Given that goal, identifying the needs of a subscriber becomes the first task. A *preliminary structure* that probably covers most user needs can be easily established from the scenarios described.

Customers want to be able to call

- A particular person, regardless of his or her location;
- A particular person at a particular location or within a particular area;
- An alternative person if the first cannot be reached;
- A particular location, rather than a particular person or a particular function;
- A particular function, rather than a particular person or a particular location;
- Any function situated within x miles of the caller.

These customer requirements of the future, which will be solved by intelligent networks, together with the ongoing development and future possibilities of intelligent network technology, suggest that in the 1990s the telephone could play a much greater role in our lives than ever imagined. The term "telephone," as it is used here, also refers to future terminals with greater built-in intelligence and with a more intelligent user interface than the poor interface available in today's ordinary DTMF pushbutton phone. An increasing number of daily functions currently handled by other media will be provided by telecommunication network services in the 90s. This will occur first through the implementation of intelligent networks in *public switched telephone networks* (PSTNs) and later in narrowband and broadband in the *integrated services digital network* (ISDN).

Today's users can get a taste of the future when they use a telephone to

- Obtain bank services,
- Obtain the daily news, stock market reports, and weather forecasts,
- Turn on the heat in their weekend cottage before arriving.

These examples point to the fact that the telephone has already begun to compete with other media, such as newspapers, letters, radio, television, and text-television. Such services are normally offered in telecommunication networks through a premium rate number, with the customer paying per service or per minute. In future, services based on broadband ISDN—for example, movies—will be offered to subscribers through the telephone network via a premium rate number controlled by an SCP in an intelligent network. Subscribers will pay for the movies via their telephone bills.

Other major areas for combined intelligent network/broadband services include interactive video services, which will allow you to sit at home and search through a large database, obtain information, and view pictures on your screen while searching. For example, from the comfort of your sofa, you could review various destinations and obtain a view of the beach, hotel restaurants, hotel rooms, and so on, before deciding where to spend your holiday. Real estate agents could use this service to permit buyers to examine different properties without leaving home. Last but not least, adventure games will probably be available through interactive video services. Intelligent network technology and plans around the world support this evolution.

What then is so special about intelligent networks? Briefly, the driving force behind introducing the intelligent network concept in networks all over the world is the goal of reducing the time required to implement a new service in the network. Conventional techniques and conventional software programming require somewhere between two to five years from the date a decision until a service becomes available to users, that is, the subscribers. The goal of intelligent networks is to reduce this time to a maximum of six months! Section 2.2 describes how that can be achieved.

2.2 THE INTELLIGENT NETWORK CONCEPT

When the intelligent network concept was first introduced in the 1980s, it was soon followed by such concepts as intelligent network services, intelligent network technology, and the like. The natural question is, did any intelligence exist in networks prior to the advent of intelligent networks?

Obviously, there has been a continuous growth of intelligence in networks since long before the term "intelligent network" was coined. Accordingly, the term "intelligent network technology" is appropriate because it describes a new type of technology, the core of which is centralized service control located in SCPs. SCPs are defined in Section 2.2.1 and are further described in Chapter 3 and subsequent chapters. The principal task of SCPs in early intelligent networks is to control the setting up of *number translating services* in the network; the most important part of the task is determining the location to which a service, for example, a freephone call, should be routed.

Services controlled by SCPs, for instance, freephone calls and credit calls, were available on the network before intelligent network technology evolved, in other words, before service control was moved from the local node (the exchange) to a centralized node (the SCP). Subscribers have been using these services for many years, and when network operators switch to intelligent network technology, by moving service control from a local (or transit) exchange to an SCP, subscribers will hopefully see no difference in use.

The obvious question then is, has the service suddenly gone from being unintelligent to intelligent? The obvious answer is, of course not. The concept of "intelligent network service" is therefore misleading. Consequently, this book uses "services" instead of "intelligent network services" when discussing *services in general*. Services can thus be implemented by using conventional, node-based technology directly in the nodes *or* by using intelligent network technology in the network *or* directly in the terminals.

This range of choices may create a situation in which different services working simultaneously are implemented in different ways. In this book, if the implementation method is of importance, services implemented using intelligent network technology are referred to as *intelligent network–based (IN-based) services*, services implemented in terminals are *terminal-based services,* and services implemented in conventional node-based ways are *node-based services*. Actually, it is quite possible that, during an updating period, for example, a network could have both a node-based and an IN-based implementation of a particular service running simultaneously for some time.

The next section defines and discusses the first standardized model for the intelligent network, the intelligent network conceptual model of the International Telecommunication Union (ITU-T) and the European Telecommunication Standardization Institute (ETSI).

2.2.1 The Intelligent Network Conceptual Model

It is essential to distinguish between *functions* and *implementations* in the intelligent network concept. It is also essential to distinguish between intelligent network architectures and the intelligent network conceptual model. While architectures will evolve with increasing service requirements and emerging technologies, the conceptual model is intended to remain consistent. It was created by ETSI and ITU-T and is described in the Q.120x series from ITU-T.[1] The goal of the conceptual model is to encourage better understanding of the intelligent network concept. The model, illustrated in Figure 2.1, is a four-plane structure, consisting of

- A service plane
- A global functional plane (GFP)
- A distributed functional plane (DFP)
- A physical plane

---

1. There is a logical structure in the numbering of the Q documents to facilitate use. The Q.120x series describes the conceptual model. Q.121x describes the corresponding standards for *Capability Set 1* (CS1), Q.122x describes *Capability Set 2* (CS2), and so on.

The following list illustrates the relationship between implementation and the theoretical model (conceptual model).

Implementation	*Conceptual Model*
a) Services	Service plane
(The plane that users see)	
b) SIBs	Global functional plane
(Service-independent building blocks)	
c) FE	Distributed functional plane
(Functional entities, e.g., CCF, SCF, SSF)	
d) Physical products	Physical plane
(E.g., SCP, SSP, SDP)	

The Service Plane

The service plane, which is described in ITU-T rec. Q.1202 [1], provides a view that is exclusively service-oriented. In other words, it shows services without indicating how they have been implemented. A service might be implemented on an intelligent network platform (IN-based) or a conventional technique might be used (node-based or terminal-based).

Service features (SF) (see Figure 2.1) are the smallest functions on this plane. A service is composed of one or more SFs. An SF is larger than a *service-independent building block* (SIB) but smaller than a service. The service plane also contains customization of services and management services.

An important matter to consider with regard to the service plane is service and service feature interaction. In other words, do undesired effects occur when two or more services or service features are used together? Service interaction, which is discussed in more detail in Section 6.4, can be divided into two parts:

1. Interactions that result from the specifications of the services (and therefore are independent of the implementation).
2. Interactions that result from how the services were implemented. (The same service, from the user's perspective, may be implemented in different ways, causing different types of interactions with other services. See Section 6.4.)

Only the first type of interaction can be solved on the service plane. The second type must include the other planes.

The SFs on the service plane are mapped to the *global functional plane* (GFP) by combining SIBs on the GFP using *global service logic* (GSL).

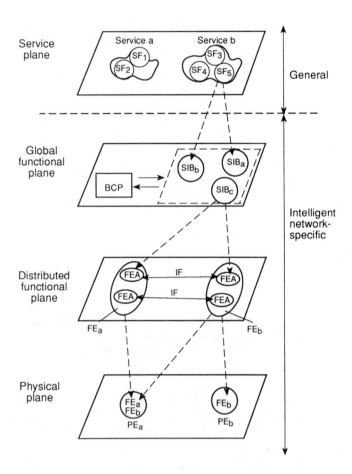

Figure 2.1 The intelligent network conceptual model. (ITU-T Rec. Q.1201, Figure 3.1/a.[2])

The Global Functional Plane

The GFP, which is described in ITU-T rec. Q.1203 [2] on architecture and in Q.1213 [3] on CS1, is the plane on which SIBs are found. GFP is also the plane on which network operators create basic new services.

SIBs are the smallest building block that can be found in an intelligent network. A SIB is defined as "a standard reusable networkwide capability residing in the global functional plane used to create service features" [2]. The *basic call process* (BCP) is a special

---

2. Full text of Q.1201 may be obtained from the ITU Sales Section, Place des Nations, CH-1211 Geneva 20, Switzerland, Tel: +41 22 730 51 11, Fax: +41 22 730 51 94.

SIB that handles all activities necessary for a normal call. Service features are composed of one or more SIBs; the SIBs are chained together by global service logic (GSL). The SIBs themselves, including BCP, are service-independent and have no knowledge of the next SIB in the chain. Consequently, GSL is the only element on the GFP that is service-dependent. The description of how the SIBs are linked together is often called a service script.

A first set of SIBs that will be commonly used, Capability Set 1 (CS1), is recommended by ITU-T and ETSI. CS1, at its current status (autumn 1993), is described in Section 2.4.4 and it is also described, together with SIBs, service features, and services, in the Appendix. Examples of service creation in CS1 are described in Section 4.2.1. There are, however, differences between the CS1 from ITU-T and the CS1 from ETSI. ETSI has, for instance, added seven extra SIBs. (Note that this discussion concerns the CS1 of autumn 1993).

CS1 will not be a standard, only a recommendation. The successors, however, CS2, CS3, and so on, are (at the time of publication) intended to be standards. The focus in subsequent *capability sets* will be on such aspects as mobility and interworking between intelligent networks and other networks.

A SIB on the global functional plane must be found in at least one *functional entity* (FE) on the *distributed functional plane* (DFP), but it can also be found in more than one FE.

The Distributed Functional Plane

The DFP, which is described in ITU-T rec. Q.1204 [4] on architecture and in Q.1214 [5] on CS1, contains FEs, which are a unique group of special functions that form a part of the total intelligent network concept. The FEs provide a detailed description and a functional specification to be applied in physical realizations. The cooperation between FEs is achieved by *information flows* (IFs). The DFP architecture is vendor- and implementation-independent.

Figure 2.2 illustrates a typical intelligent network configuration today, including the interworking between FEs. Bear in mind that there is only one *service management function* (SMF), because a network can only contain a single *master*. It controls a number of *service control functions* (SCFs) and a larger number of *service switching functions* (SSFs) in the network. But, as Section 7.6 discusses, we may have to consider cooperation between SMFs in future, for instance, when mixing traffic from different network operators.

The FEs on the DFP, shown in Figure 2.2, are

- *Call control agent function* (CCAF). This function handles user access to the services. The CCAF can be found in exchanges with subscribers, that is, local exchanges.
- *Call control function* (CCF). This function handles all normal calls and the switching of calls and services, both IN-based and non-IN-based. The CCF can also recognize a service that is going to be handled by the intelligent network.

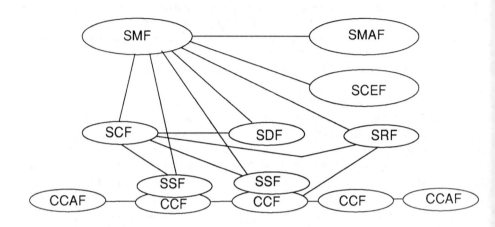

Figure 2.2 The distributed functional plane. (ITU-T Rec. Q.1211, Figure 2/Q.1211.[3])

- *Service switching function* (SSF). This function handles communication between the CCF and the SCF. The SSF is also the function for switching a call or service to a particular location (particular number) on order from the SCF.
- *Service control function* (SCF). This is the core of the intelligent network. This function controls the whole process of handling calls or services by giving instructions to the SSF/CCF, the SDF, and the SRF. Knowledge of the specific data and logic of a service itself is only found in the SCF.
- *Service data function* (SDF). This function assists the SCF, providing data about customers and the network.
- *Specialized resource function* (SRF). This function is used to give and receive data to and from users, much like control facilities for code receivers and speech machines.
- *Service creation environment function* (SCEF). This function handles defining, developing, and testing of IN-based services.
- *Service management agent function* (SMAF). This function handles the human interface with the SMF for, for example, operation and maintenance staff.
- *Service management function* (SMF). This function operates like a spider at the center of web, managing the distribution, control, and maintenance of IN-based services.

The FEs on the DFP are allocated to *physical entities* (PEs) on the physical plane.

---

3. The full text of Q.1211 may be obtained from the ITU Sales Section, Place des Nations, CH-1211 Geneva 20, Switzerland, Tel: +41 22 730 51 11, Fax: +41 22 730 51 94.

The Physical Plane

The physical plane, which is described in ITU-T rec. Q.1205 [6] on architecture and in Q.1215 [7] on CS1, shows how the aforementioned functions on the DFP can be implemented in physical products on the network, that is, how an FE on the DFP is allocated to a physical entity (PE), or physical node.

An FE must be found in at least one PE, but a PE can consist of more than one FE. (As an example, the PE service switching and control point (SSCP) consists of two FEs, the SCF and the SSF.) However, there cannot be more than one FE of the same type in one PE. In other words, according to the theoretical model, there can not be two SCFs in one SCP. If for some reason you still want to locate two SCFs in the same place, they must, from a theoretical point of view, be regarded as two different PEs, that is, two SCPs.

Moreover, it is not possible to divide an FE between two physical nodes. (Implementation aspects of PEs are discussed below and in Chapters 3 and 7.) Examples of physical nodes on this plane include:

- *Service switching point* (SSP). This node contains the SSF and call control function (CCF), recognizes an IN-based service, and communicates with the node containing the SCF (often the SCP). If the SSP is a local exchange to which the users are connected, the CCAF is also performed. The SRF may be integrated in the SSP or directly in an IP.
- *Service control point* (SCP). This node contains the SCF, namely, the service logic that controls the IN-based services.
- *Service switching and control point* (SSCP). This is a physical node that contains both the SSF and the SCF.
- *Service management point* (SMP). This node is used for operation and maintenance tasks, such as provision of services, collection of statistics, and so on, which together form the service management function (SMF).
- *Service data point* (SDP). This node contains the SDF, that is, data used by the service script. An SDP may be accessed from either the SCP or the SMP.
- *Intelligent peripheral* (IP). This node contains the *specialized resource function* (SRF) and is used to communicate with the users of a service. It can be used for voice recognition, code receiving, protocol conversion, customized messages, and so on, and for communicating with one or more SSPs. The SRF might as well be integrated directly in the SSP.
- *Adjunct* (AD). This node is similar to an SCP, but has a high-speed interface to the SSP.
- *Service node* (SN). This node is similar to an SCP and an AD, but has a direct point-to-point connection with the SSP for speech and signaling.
- *Service creation environment point* (SCEP). This node contains the SCEF and is used for defining, developing, and testing IN-based services.

- *Service management agent point* (SMAP). This node contains the SMAF and is used to access the SMP. Access can be given to the network operation staff, as well as to external, specially selected customers.

Implementation Aspects

The most common way to implement the SCF and SSF is in the SCP and SSP, respectively. The SCF, for instance, is the control intelligence of services such as freephone. In the early stage, before the first SCP was installed, many networks implemented the SCF for number translation services, such as freephone in a transit exchange. (This is a very common way to start offering these services in a network.) When intelligent network technology was introduced and the first SCP was implemented, the most natural evolution was to move the functionality represented by the SCF to the SCP, along with control of the freephone service.

The SSF is the function that switches the call upon receipt of orders from the SCF. The normal way to implement an intelligent network is to separate geographically the SCF and the SSF and to have one or more SCPs (with the SCF) control an entire network of SSPs for a service. But, to do this, a signaling possibility between the SSPs and the SCP is needed, that is, a signaling system with both high capacity and a flexibility that also allows communication of information other than normal call setup data. In other words, what is needed is a *data communication channel* separated from the speech channels. *Common channel signaling* (CCS) provides that facility, and consequently, CCS is a fundamental base for building the required signaling within an intelligent network. In the *Open System Interconnection* (OSI) application layer (layer 7) in *Common Channel Signaling System No. 7* (CCSS No. 7), which is the most common CCS system today, a protocol, *Transaction Capability Application Part* (TCAP) is introduced to provide the intelligent network with the signaling required. Within TCAP there is a component layer, *Intelligent Network Application Protocol* (INAP) that is used for intelligent network communication, for example, between SSPs and SCPs, between SCP and SDP, and so on. See Section 3.4 and Figure 3.2 for more information about CCSS No. 7 and the TCAP/INAP.

Of course, this provides incentive for the INAP, based on the CCSS No. 7 signaling system, to be available in both the SCP and the SSP. It will, however, be some time before INAP is implemented in many networks; in the meantime, the following approaches can and have been used. Telecommunication networks that contain a separate signaling network, that is, a CCSS No. 7 network but where INAP is not implemented yet—as in the first years of intelligent networks in Sweden—can still physically separate the SSF and the SCF prior to implementing the INAP. This can be done by using the facilities in the CCSS No. 7 to make speech and signaling (quasiassociated or nonassociated signaling) take different paths in the network. (The technique is described further in Chapter 3, Section 3.4.) For networks in which CCS is not implemented, the best way to implement an intelligent network is to allow the SCF and the SSF to be physically present in the same node, that is, in an SSCP. This way, a signaling protocol is not required; the communication between the SCF and the SSF is performed via an internal interface within the software in the SSCP instead.

The advantage of these last two approaches is that IN-based services can be made available earlier, instead of having to wait for the INAP to be implemented, which is frequently one of the network components for which we must wait the longest. This is because it also influences the nodes (the SSPs) and therefore must be implemented both in one or two SCPs and in a large number of SSPs.

The disadvantage of the SSCP method of implementing an intelligent network is that we must also switch the speech channel to the location of the SCF, that is, to the SSCP. This slows down the capacity of many network services, since the possibility of having a free speech channel between a node (local or transit) and the SSCP determines whether you can reach the SCF or not. This could be especially devastating for mass calling services, such as televoting, as well as for premium rate services. (Refer to Section 3.4 for a description of SSCP and to Section 3.5 for a discussion of mass calling services).

One advantage of the intelligent network is, as described above, that it allows a service to cover an entire network and be controlled from one central point (an SCP) in the network instead of from many (local) exchanges. This facilitates such administrative and management procedures as introduction, change, and withdrawal of services. Centralized control and centralized administration of a service also provide the operator with a better overview of that service in the network. As it becomes easier to introduce new services by using the intelligent network concept, a much wider range of services will undoubtedly become available.

2.2.2 The Introduction of Platforms

The goal of the intelligent network concept is, as described in Section 2.1, to reduce the time required to implement a new service from between two to five years , using conventional technology, to less than six months, using intelligent network technology. How is this possible? It becomes possible by taking a completely new approach to the whole service implementation complex—by building the intelligent network using a *platform* concept (see Figure 2.3).

An advantage of this method is that all the software commonly required for a range of services is already included on a stable platform that seldom changes. Consequently, when a new service must be designed and programmed, only the parts that are unique to the service need to be developed and placed on the platform to make the service ready. Furthermore, the unique service parts are based on supersophisticated programming by means of service-independent building blocks (SIBs). The "programming" of a new service will therefore differ considerably from conventional programming. A suitable range of SIBs will be put together to form a service, and, once appropriate parameters have been set, the service will be ready. In future, using the intelligent network concept, a new service will be introduced in months or perhaps even weeks, provided all the SIBs needed actually exist. If a SIB is missing and must be designed, it will necessarily take longer to introduce a service. But the probability that all the SIBs required for a new service on a

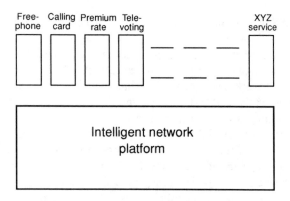

Figure 2.3 The intelligent network employs a platform concept.

major intelligent network will actually exist should in future be closer to 100% than to 90% because the growing number of services on the intelligent network will result in a growing library of SIBs.

2.2.3 What the Intelligent Network Really Is

Phase One: Separation of Switching and Intelligence

The introduction of intelligent networks represents something of a revolution in that we are starting to separate the intelligence—that is, the logic to set up a call or to control a service in the network—from the switching. Intelligence and switching functions have historically been very closely integrated in the telephone exchange, with no possibility of separating them. A call is set up according to a step-by-step procedure, with each node deciding which node to proceed to next, making it impossible to obtain a total overview of the call setup from one location. The situation is the same for setting up a service, like freephone, through the network. See Figure 2.4.

When the control (the SCF) of a service, such as a number translation service, like freephone or premium rate, is moved from the nodes to an SCP, control of the whole service is centralized at a single point on the network. This point, which is the SCP, has a total overview of the execution of the service, including logic, data, and management. If the service is changed or withdrawn from the network, or if a new service is introduced, it is mainly the SCP that is affected. Of course, we have to "teach" the SSPs to call the SCP upon recognizing an indicator that a service must be activated and then to wait for orders from the SCP. This can be described as a *liberation* of intelligence from switching in the networks. See Figure 2.5.

Along with this liberation, it also becomes possible to put intelligence in more than one place on the network. Provided there is good and secure network signaling, we can distribute intelligence, that is, the control facility of a service, anywhere on the network.

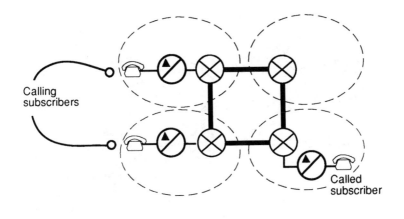

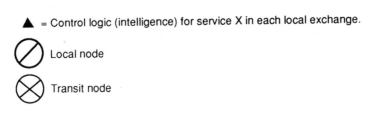

▲ = Control logic (intelligence) for service X in each local exchange.

⊘ Local node

⊗ Transit node

Figure 2.4 Example of service setup before intelligent networks: node-based control. (Service X can be freephone services, call forwarding unconditional, and so on.)

The discussions so far provide an accurate picture of where the intelligent network concept is heading in future, namely, towards the possibility of *permitting a more flexible allocation of the service control function* than ever before.

Phase Two: Flexible Intelligence Allocation in the Network

The flexible allocation of the SCF in the network is the second phase of intelligent network evolution (see Figure 2.6.), whereby the intelligent network permits control of a service that was initially fixed to a local node, where the subscriber was connected, to be carried out by any physical point in the network.

Let us leave, for a moment, the theoretical discussions and consider the reality in many networks today, and we will realize that phase two is already underway. There are three major trends in service logic allocation:

1. Services for which service logic (the SCF) may be moved from the local or dedicated service node to the *terminal* (the phone). Examples include:
   - Repetition of last dialed number
   - Abbreviated dialing

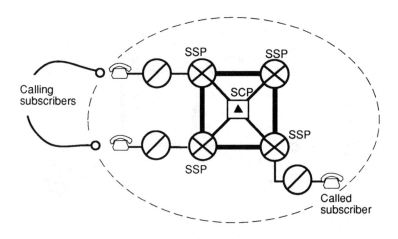

 = Control logic (intelligence) for service X centralized in *one* SCP

⊘ Local node

⊗ Transit node

▲ Service control point

Figure 2.5 Service setup in phase one of intelligent network evolution: separation (liberation) of intelligence. (Service X can be freephone services, call forwarding unconditional, and so on.)

2. Services for which service logic (the SCF) may be moved from the local or dedicated service node out into the network, that is, to SCPs. Examples include:
   • Freephone
   • Premium rate
   • Call forwarding unconditional
   • Call transfer services
3. Services for which service logic (the SCF) will *remain in the local or dedicated service node* because it is very hard to move. Often it is because these services require real-time communication between nodes, that is, advanced real-time signaling. This does not, of course, mean that the SCF for those services can not be moved to other places later on. Examples include:
   • Call completion on busy subscriber
   • Call completion on no reply

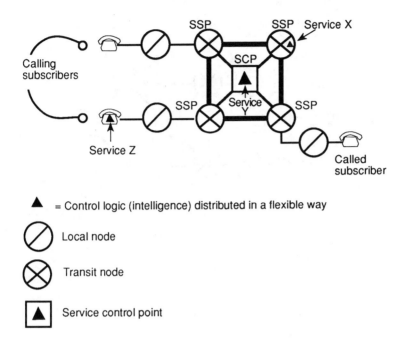

= Control logic (intelligence) distributed in a flexible way

Local node

Transit node

Service control point

Figure 2.6 Phase two of intelligent network evolution: flexible intelligence allocation. The *right* service in the *right* place.

By placing the logic in the terminal, we no longer require signaling in the network; the terminal handles the service autonomously. The advantages of placing the logic in an SCP (intelligent network) are that

- We make the service available to *all* subscribers at once.
- We gain a total overview of the service from a single location on the network, which facilitates introduction, removal, maintenance, and the like.

Phase Three: Intelligence on Demand

Farther into the future, we will most likely be able to retrieve logic or intelligence wherever we need it (see Figure 2.7). This is the third phase. Intelligence on demand is discussed further in Chapter 7, Section 7.5.2. (Resource optimization and allocation are covered in more detail in Section 7.5.)

With the introduction of the intelligent network and its SCPs, we have taken the first step towards a future in which a global intelligence, completely separate from the network, controls call setups and services. Intelligence may be retrieved wherever users need it, and

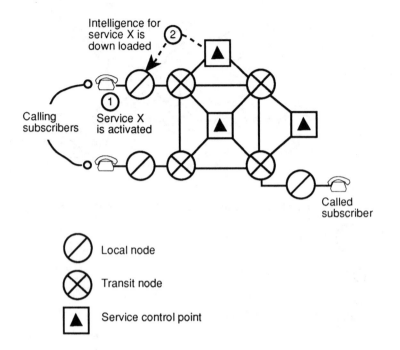

Figure 2.7 Phase three of intelligent network evolution: intelligence on demand. When the subscriber activates service X, the control logic (intelligence) is downloaded to the local node. The *right* service in the *right* place at the *right* time.

all activities will be optimized because a call path through the network will be set up, based on *minimum cost and delay*. In future, the minimum cost call setup will depend on, among other factors, the shortest path, momentary load situations, and the priority level of the user.

2.2.4 Applicability to All Access Networks

The intelligent network as a concept is general, a level above access networks, such as the PSTN, ISDN (narrowband and broadband), cellular networks, and so on. (See Figure 2.8.) Compared to an intelligent network, these access networks are reduced to access forms. Consequently, in the future it should be possible to set up freephone calls, premium rate calls, credit calls, and so on regardless of the type of network used to access the intelligent network level. The intelligent network level should be able to offer the same capabilities (i.e., services) to all access networks and in the same way, for example, making number translating dependent on time and origin, charging (if desirable), providing a customer control interface, and so on.

Furthermore, it should be possible for users with different means of access to call one another, depending, of course, on the condition of the weakest party and on the rele-

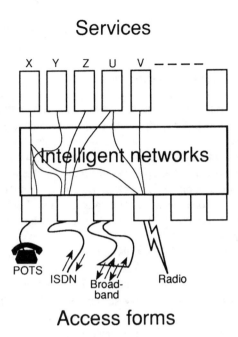

Figure 2.8 The intelligent network is a general concept.

vant services. The intelligent network will control the call setup and might also determine whether the connection between two users with different means of access is allowed for the call or service (see Figure 2.9).

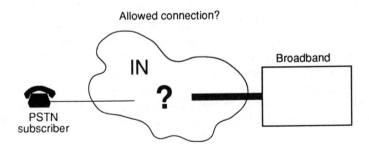

Figure 2.9 The weakest party determines the conditions.

2.3 NETWORK SERVICE IMPLEMENTATION

This section discusses network services and how different implementation methods influence how services are accomplished, the time required, and the degree of flexibility that can be achieved. Both the intelligent network method and conventional methods of implementation are examined.

2.3.1 Conventional Service Implementation

The conventional method of introducing services is, as everyone knows, a rather time-consuming and complicated task. The introduction of a new network service using conventional techniques (conventional programming, testing, and integration in every network node) requires between two and five years in an ordinary network today. The requirement to reduce this time drastically is the principal driving force behind the intelligent network concept, as discussed in Section 2.1. A lack of flexibility characterizes the service packages available in a conventional network. The packages are often highly inflexible, stable, and limited in number of services. Moreover, these services have usually been available to users for quite some time.

This last observation is especially true, as many of the services introduced in the network by conventional techniques were available in PABXs long before they were introduced into public networks. Examples include call forwarding unconditional (CFU), call transfer (CT), and call completion to busy subscriber (CCBS), all of which were implemented in PABXs in the 1970s before becoming a reality in some public networks in the 1980s. Consequently, subscribers who have been using both PABXs and the public network have been using these services for at least a decade.

The number of services implemented by conventional means in public networks is normally somewhere around 20 to 30 and will probably never exceed 50 in a particular network. The way in which a specific service is implemented, which, from the subscriber's point of view, could affect the way the service behaves, often differs from network to network. Nevertheless, a certain network usually never uses more than one method of service implementation for a single service if conventional techniques are used.

Furthermore, in a particular network, it is extremely difficult to find different features or variants of a particular service adapted to different customer (subscriber) needs. This means that each subscriber, or user, in a network has exactly the same type of service. However, when comparing different networks, the same service may have been implemented using a different method and may be functioning differently. Often, these minor differences in service implementation are disastrous when a service is used by users in different networks. Call completion to busy subscriber (CCBS) is an excellent example of a service that, at the time of publication, is very difficult to get to work outside your own system.

2.3.2 Service Implementation on Intelligent Networks

The main goals of introducing services using the intelligent network concept are to solve the problems and eliminate the inflexibility described in Section 2.3.1 and to reduce the time required to implement a new service to less than six months—from the time the decision is made until the service is available to subscribers. This can be accomplished by using a stable platform that contains common functionality for many services. The unique parts of a particular service are placed on top of the platform, creating a complete service for the users. Only a minor part of the service has to be developed. Furthermore, the unique parts are built using the SIB technique. (See Section 2.2.1.) From a library of SIBs, which are large software building blocks, a suitable number of SIBs are chosen to form a service. (See Figure 2.10.) Once the necessary parameters are established, the service is placed on the platform, tested, and, finally, offered to the subscribers. (See Figures 2.11 and 2.12.) Consequently, the testing procedure is rather short. But there is another type of testing that will become absolutely essential in future, namely, testing of the interaction between a new service and existing services. (See Section 6.4.)

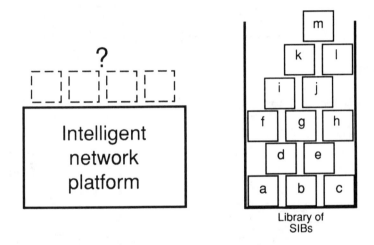

Figure 2.10 The method for building services in intelligent networks.

Service-Independent Building Blocks

Section 2.4.4 and the Appendix describe the SIBs in CS1, from ITU-T. In future, CS1, or at least the same functionality as CS1, will probably be implemented in all intelligent networks; however, as CS1 was not ready nor stable at the time intelligent networks became available, the SIBs in CS1 are not exactly the same as the SIBs used in networks today. Today, SIBs are, of course, highly influenced by the products offered by different intelligent network vendors. Although the basic principles for creating services, using platforms, and offering SIBs are the same for all vendors, no two vendors offer identically functioning SIBs. Other differences between CS1 SIBs and the SIBs offered by vendors are the size and, consequently, the number of SIBs needed to create a particular service. In other words, different intelligent network implementations use different levels of SIBs; some use a few large SIBs, while others use a large number of smaller SIBs to achieve the same

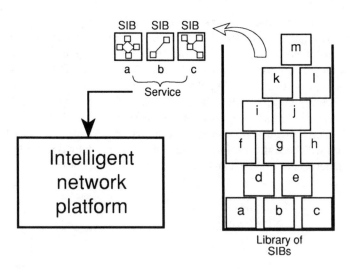

Figure 2.11 SIBs are combined to form a service.

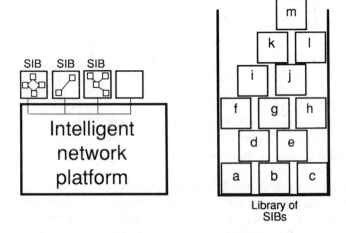

Figure 2.12 The service is placed on the platform.

service functionality. The general method of using basic software blocks is, however, the same for all intelligent network implementations.

Typical functionality in intelligent network implementations (driven by real needs) includes SIBs that handle

- Number analysis

- Time-dependent translation
- Call origin–dependent translation
- Percentage distribution of calls to different destinations
- Queuing possibility

Combining SIBs into Services

SIBs are combined in a particular way to form a complete service. (See Figures 2.10, 2.11, and 2.12.) When the appropriate SIBs have been combined, only a few parameters need to be established to make a service ready. If the appropriate SIBs exist and the person creating the service is experienced, the procedure can be accomplished in about one month. (Section 4.2.1 describes how to build services for intelligent networks in more detail and provides examples.)

Service Provisioning and Customization

Once a service is created, it is considered a *basic version* of the desired service. But different users of the service have different requirements; therefore, the service has to be adapted to meet the special needs of each customer. This is called service provisioning and customization. Consider, for example, how a pizzeria delivery service that has multiple locations and thus needs to locate the origin of calls uses the freephone service. Here, the network operator and the pizzeria management will have to work together to divide delivery service into geographic areas. The divisions can also be time-dependent, that is, customer calls can be routed to different locations depending on the time the calls are placed. (Section 4.2.2 describes provisioning and customization of services and provides more detailed examples.)

Service Implementation Demands an Overall Approach

The aforementioned IN-based service is implemented on the intelligent network platform on the network (in an SCP) and can then be reached from many SSPs. This is a new approach to the technical aspects of installing services whereby, primarily, the time needed to design a service is drastically reduced. But there are numerous other matters that must be taken care of before a service can be made available to subscribers. Therefore, to really reduce the introduction time for services, it is necessary to take an *overall approach to introducing an intelligent network.* The technical platform described above is only one factor and will not be worth anything unless the other systems in the network—management systems, charging and billing systems, and so on—can be updated in the same short time. Accordingly, an overall approach to the whole chain of systems must be considered in the rapid introduction of services on a network. (See Figure 2.13.) (This is further discussed in Section 6.3.)

The software design period, which, prior to development of the intelligent network concept, was the major component of introducing new services, will become the shortest

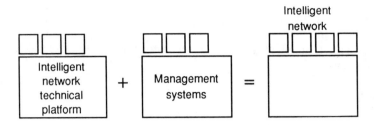

Figure 2.13 Introducing intelligent networks requires an overall approach.

part. Instead, introduction time of a service will be determined in future by the ability of the surrounding systems to adapt to the IN-based way of building services. Consider the following:

- An organization decides to implement a new, conventional service as soon as possible. Software design will take 18 months, testing and integration, six months; in total, it will take 24 months from start until the service can be used on the network.
- The start time for software design is also the start signal for preparation of the management systems that must be ready when the service is put into operation 24 months later. Management system preparation takes 12 months, so there is no hurry to complete that task compared to the 24 months needed for technical development.
- When the intelligent network is introduced, technical development is reduced from 24 to six months, a reduction of 18 months.
- But development of the management system still takes 12 months! To avoid a bottleneck, management system development must also adopt an IN-based way of working that will reduce development time to six months or less.

Withdrawal of an Undesired Service

Organizations must also be able to close or withdraw an undesired service quickly from a network. This is an often-overlooked but highly important matter, usually necessitated by a severe situation in the network, for example, misuse of a service, problems with service interaction, or charging problems. The activities around a withdrawal must also include closure of the service in the management systems.

2.4 THE EVOLUTION OF INTELLIGENT NETWORKS

2.4.1 How It All Began

Intelligent networks emanated from the first introduction of the 800 service (freephone) in the United States in the 1980s. The term "intelligent network" was introduced by BellCore around 1984 and signaled the start of a new era in telecommunication. At this time it became clear that centralized intelligence for controlling services nationwide could provide such extensive benefits that it was worthwhile to build a separate node *service control point* (SCP) exclusively for this purpose.

2.4.2 Bellcore's IN/1, IN/2

So, when intelligent network was conceived, planning for its future began. BellCore was first on the scene again, defining two concepts, IN/1 and IN/2, which, if implemented in order, would gradually increase the intelligence in a network.

IN/1, the easiest to implement, was a straightforward concept for introducing number-translating services like freephone, that is, services for which the SCP was a simple "translator" from one telephone number to another. Concepts similar to IN/1 have been implemented in most countries today and represent a natural first step toward implementing intelligent networks.

With the IN/1 concept, when a new service was introduced, *both* the SSP and the SCP had to be updated, a fact that was seen as a drawback for the future. Consequently, it was decided that in the IN/2 concept *only* the SCP would be updated for a new service. The idea was to simply download the software for a service from the SCP to the SSP via the network. Very soon, however, it became evident that the IN/2 concept was too difficult to implement in one step, so an interim concept was discussed (IN/1+). This was also later abandoned and instead focus became fixed on a new concept, *advanced IN* (AIN).

2.4.3 Advanced IN

The main idea of AIN, compared to its predecessors, is to be *service-, switch-, and equipment-independent*.

In 1991, BellCore presented a technical advisory on AIN Release 1. It was primarily meant to be a target document; a direct step to AIN Release 1 would be impossible. The result, however, has been that CS1, from ITU-T/ETSI, has become more or less a subset of BellCore's AIN Release 1.

2.4.4 Capability Sets from ITU-T and ETSI

In the standardization of the intelligent network concept in ITU-T and ETSI, work has been divided into a number of concepts, the first ones being Capability Set 1 (CS1), Capability Set 2 (CS2), Capability Set 3 (CS3), and so on.[4] The term "capability set" refers to the set of services and service features that can be constructed by using the SIBs contained in each concept.

---

4. Note that this discussion relates to the status of intelligent network technology in autumn 1993.

There are, however, distinctions between the capability sets from ITU-T and those from ETSI. For CS1, ITU-T has defined 13 SIBs plus a SIB for the *basic call process* (BCP). ETSI has defined the same SIBs plus seven more to create an ETSI CS1. The CS1 from ITU-T is described briefly here and in more detail in the Appendix. Examples of service creation are described in Section 4.2.1.

CS1 was formulated as a standard, but will only be a recommendation. CS2 and CS3, on the other hand, are aimed at being standards, and their focus is on such items as mobility and interworking between intelligent networks and other networks. The main idea is to standardize interfaces and SIBs, but not the services themselves. However, to facilitate understanding of how to use the SIBs, CS1 also describes a set of service features and services that can be built using its SIBs. These services represent *a minimum set of services* that can be built with the help of the SIBs in CS1 and may be particularly useful when an intelligent network implementation is being tested for conformance to a capability set. (Conformance testing determines whether a system meets the requirements of a certain standard.) In future, it will, of course, be possible to build more than these services; this is also a main goal of the intelligent network concept.

In CS1, the following limitations are imposed:

- Services should be *single-ended*, that is, the service should concern only one user.
- Services should have a *single point of control*, that is, only one service control function may be involved, and an SSP may have a relationship with only one SCP. Direct communication between two SCFs is not permitted in CS1.
- Services defined by CS1 will support the networks PSTN, ISDN, and PLMN.
- The only interfaces defined in CS1 are between:
    □ The service control function (SCF) and the service switching function (SSF);
    □ The service control function (SCF) and the specialized resource function (SRF);
    □ The service control function (SCF) and the service data function(SDF).

The SIBs defined in CS1 from ITU-T (Rec. Q.1213) [3] have the following functions. (The SIBs in CS1 and the minimum set of services they support are described in detail in the Appendix; examples of service creation are given in Section 4.2.1.)

CS1 SIB Functions	*Description*
Algorithm	For mathematical algorithms
Charge	To influence charging
Compare	To compare two values
Distribution	Call distribution to different destinations
Limit	To limit the number of calls for a service
Log call information	To log detailed information about a call
Queue	Sequencing of calls to a called party
Screen	To check if a number is in a specified list
Service data management	Management of end user data

CS1 SIB Functions	Description
Status notification	To check the status of a service
Translate	To translate input information into output information
User interaction	To exchange information between the network and involved parties
Verify	To compare collected information with expected format
Basic call process	A specialized SIB, providing the basic call capabilities

The following list contains the minimum set of services that can be built using the SIBs in CS1. (A functional description of each SIB and the services it supports is provided in the Appendix.) The services are, consequently, *fully* supported by the SIBs in CS1 from ITU-T (Rec. Q.1211) [8]:

CS1 Minimum Services	Description
Abbreviated dialing (ABD)	Use of short numbers
Account card calling (ACC)*	Call from any terminal, charging an account
Automatic alternative billing (AAB)*	Billing via a separate account
Call distribution (CD)	Routing incoming calls to different destination(s)
Call forwarding (CF)	Forwarding of incoming calls to another number
Call rerouting distribution (CRD)	Rerouting of incoming calls due to a condition of busy, after a certain number of rings
Credit card calling (CCC)*	Calls from any access points charged to a specified account
Destination call routing (DCR)	Routing of incoming calls depending on time, date, originating area, calling line, etc.
Follow-me diversion (FMD)	Remote control redirection of incoming calls
Freephone (FPH)	Reverse charging
Malicious call identification (MCI)	Logging of incoming calls
Mass calling (MAS)	Preparation for a mass calling service
Originating call screening (OCS)	Restriction of outgoing calls
Premium rate (PRM)	Partial pay back for a call to the called party
Security screening (SEC)	Security screening of users seeking access to network, system, or applications
Selective call forwarding on busy/no answer (SCF)	If called party is busy or does not answer, only particular, preselected calls may be forwarded
Split charging (SPL)	Split charges between calling and called parties
Televoting (VOT)	Preparation for a televoting session
Terminating call screening (TCS)	Restriction of incoming calls
Universal access number (UAN)	Using one number for several incoming lines
Universal personal telecommunications (UPT)	Using one unique number across multiple networks at any access

(continued)

User-defined routing (UDR)	The user may specify routing of outgoing calls
Virtual private network (VPN)	Building a private network using public network resources

*These services, initially different from each other, are very similar today. CCC, for instance, can also be used without a card. All three, ACC, AAB, and CCC have the common characteristic that a call may be charged to an account.

The following two services are only *partly* supported by the SIBs in CS1:

Partly SIB-Supported Functions	*Description*
Call completion to busy subscriber (CCBS)	Calling party is automatically informed when called party becomes free
Conference calling (CON)	Multiple parties participate in one single conversation

Standardization efforts are under way for CS2, but the SIBs and services have not been defined yet (autumn 1993).

2.5 INTELLIGENT NETWORK USERS: REQUIREMENTS AND OBLIGATIONS

A general requirement of services in an intelligent network is that they should offer at least the same degree of availability to the user as ordinary telephone traffic, which is quite high in most countries today, at least compared to many other systems used in daily life. (Regarding future user needs, please refer to Chapter 7.)

2.5.1 Service Subscribers and Service Users

Service subscribers and service users are the first groups who come to mind when intelligent network users are considered. The distinction between the two is that service users use the services and service subscribers own the subscription (pay the bills). Often they are the same people, but not always. Within a family or a company, for instance, there is often one person who owns the subscription and several others who use the services.

Service users reach services via routines managed and offered by the network, that is, via their terminals (telephones). The services themselves are provided by service providers, who can be network operators or external providers offering service via the network. Users want services well-suited to their private and/or professional lives. They want good, easy-to-use, and stable services. These same requirements apply to the user interface. A "stable" service is one that functions as close to 100% of the time as possible and does not constantly change due to the operating sequence.

An example of a demand that would come from service subscribers is specification of the calls and services to accompany a bill. (For many years, specified bills have been a reality for many subscribers in the United States, but they are still at an introductory stage in Europe.) A customer control facility is a future demand for both service subscribers and service users. (For a description of customer control facilities, refer to Section 2.5.2, Section 4.3.2, and Section 6.3.2.)

2.5.2 Service Providers

Service providers, who offer services to the customers of a network operator, are another kind of user of intelligent networks. These services may be manual services offered by a person, such as an attorney or a physician, *or* automatic services offered by a machine, such as news, stock exchange reports, and weather forecasts. Service provider requirements for availability, that is, the ability to offer uninterrupted service, are even higher than for normal subscribers. This is usually because the network often provides the only opportunity service providers have to sell their services.

Customer control is an important feature for service providers and for service subscribers. Among other things, customer control allows the C number (the result from the number translation in the SCP) to be changed; in other words, the answering location can be changed. For example, a service provider who wants to change the routing of incoming calls, for example, because of different times or different locations, can, from his own terminal, call the service management point (SMP) and, after a security check, reach into and change his own data. The SMP then updates the SCP. This feature is described in more detail in Sections 4.3.2 and 6.3.2, which discuss the future development of customer control functionality. Of course, today a user friendly interface to the customer control function is not only desirable, it is a strict necessity.

2.5.3 Network Operators

We tend to overlook the fact that network operators are also intelligent network users. Many of the services in a network are only used by network operators, for example, operation, administration, and maintenance services. Network operators are, of course, responsible for establishing connections between other users, but they are also responsible for financial transactions between service subscribers and service providers for premium rate services, televoting services, and so on.

Second Network Operators

Second network operators represent another user category. These operators pass a call or service from their network to another network. If, when it reaches the other network, the call or service is then controlled by that network's intelligent network, the setup through that network must be assisted by its SCPs.

In future, the customer control function for routing calls through a network may also be available to other network operators passing the network.

2.6 THE TERM "INTELLIGENT NETWORK" WILL NOT BE USED BEYOND THE 1990s

The buzzword on everyone's lips in the telecommunication area in the 1970s was "SPC," which was a new concept for building telephone exchanges. With SPC, computers were used to *control and centralize intelligence in the telephone exchange.* SPC exchanges replaced the old electromechanical systems, in which the call setup within the node itself was carried out using a step-by-step procedure. Suddenly, with the SPC technique, it was possible to gain a total overview of a call setup through an exchange. But SPC exchanges are seldom mentioned today because, having largely replaced electromechanical exchanges, they are quite commonplace.

In much the same way that the SPC technique spawned a revolution in building nodes, intelligent network implementation, in its initial phase, signals the start of a revolution in building networks. (intelligent networks actually represent much more than that, as other parts of this book point out.) The first generation of SCPs on intelligent networks are simply components that *control and centralize intelligence in the network*, instead of performing the step-by-step call setup procedure in practice before intelligent networks. Consequently, by the end of the 1990s, the term "intelligent network" will no longer be used, the concept will be firmly integrated and commonplace in all networks. This will largely be a result of the centralized control capability of intelligence, and, more importantly, of further development of the network building concept, as well as of the service creation and service implementation concepts. (These are discussed in detail in Section 7.5.)

This last conclusion, that is, that intelligent networks offer an approach to the whole technical concept, combined with the fact that intelligent networks may be introduced in all access networks, suggests that intelligent networks will be *the next natural step in the evolution of any network,* in much the same way that SPC has become the only way to build nodes today. After all, who wants to return to the old step-by-step procedure of building nodes?

So intelligent networks are a principal goal of the telecommunication and data communication evolution and will sooner or later become a reality in all networks. The way this goal is attained, however, will differ from network to network, because networks will work towards intelligent networks from different starting positions. The ultimate goal, however, will be similar in all networks, in the same way that implementation of SPC exchanges has become a common way of building nodes.

REFERENCES

[1]   ITU Recommendation Q.1202, 1993.
[2]   ITU-T Recommendation Q.1203, 1993.

[3]   ITU-T Recommendation Q.1213, 1993.

[4]   ITU-T Recommendation Q.1204, 1993.

[5]   ITU-T Recommendation Q.1214, 1993.

[6]   ITU-T Recommendation Q.1205, 1993.

[7]   ITU-T Recommendation Q.1215, 1993.

[8]   ITU-T Recommendation Q.1211, 1993.

Chapter 3

Platforms for Intelligent Networks Today

3.1 GENERAL TRENDS

A number of general trends within all networks are influencing evolution. These changes, which have been gradual, include:

- Reduction in transmission costs,
- Higher bandwith availability,
- Higher network signaling capacity,
- Growth of network intelligence,
- Greater freedom in allocation of intelligence in the network,
- Greater awareness of redundancy aspects,
- Greater awareness of security aspects.

3.2 A BASIC TELEPHONE NETWORK IS A VALUABLE PLATFORM

A good *basic telephone network* provides an excellent base for implementing a good intelligent network platform in a network. But how does one build a good basic telephone network for an intelligent network platform?

We must first define what we mean by intelligent network platform. This book takes a general approach to the discussion of what constitutes an intelligent network platform. First, of course, there is the technical platform, which contains such components as local and transit nodes, transmission systems, access networks, service control points (SCPs), service switching points (SSPs), and so on. But it is also necessary to include in our definition the administrative and management platforms that are required to fulfill the main goals of an intelligent network, namely, rapid implementation of new services.

Returning to the question of how to build a good basic telephone network for an intelligent network platform, actually, a basic telephone network is not something that we build. On the contrary, it is something that we, more or less, *already have*. Building (or changing) a basic telephone network is enormously difficult, very expensive, and time-consuming. So, you will have to live with the basic telephone network you already have, at least in the near future, which, in fact, corresponds to the timeframe for building the first intelligent network platform. Of course, you can make minor improvements to your telephone network, but major changes, such as introducing a networkwide Common Channel Signaling System (CCSS) No. 7, take a very long time. Consequently, you have to find an intelligent network solution that makes the best use of your existing network and "excuses" the parts that are not quite up to date.

The following sections discuss different approaches to implementing an intelligent network, based on the basic telephone networks from which you can start. The discussion does not cover every aspect and detail of implementing an intelligent network, but the main components and the main system functions in a network are considered from the perspective of an intelligent network implementation.

3.3 VALUABLE FUNCTIONS IN AN EXISTING NETWORK

The tasks of building, changing, or expanding an intelligent network platform are facilitated if the existing network possesses any of the following characteristics:

- A CCSS No. 7;
- An automatic message accounting (AMA) record technique (toll ticketing) charging system;
- Standardized interfaces;
- A single vendor situation (i.e., SSPs and SCPs from the same vendor);
- Stored program control (SPC) exchanges, both local and transit;
- Digital transmission and Pulse code modulation (PCM);
- Dual tone multifrequency (DTMF) interfaces widely available to the subscribers (i.e., pushbutton terminals instead of older dial pulse (DP) telephones);
- Calling line identification capabilities;
- A well-established organization for marketing and managing value-added services (VASs).

These are probably the most important characteristics. You can manage without all of them, but the more you have, the more rapidly and inexpensively you can implement a good intelligent network platform and start to provide IN-based services.

A CCSS No. 7 is a valuable component to have in your network because it provides a 64-Kbps "data communication channel" for almost free use, which is necessary for, for example, handling communication between SSPs and SCPs. The AMA record technique provides the capability to move all charging activities outside the network (outside the technical system), which affords a greater flexibility in the tasks of setting and changing tariffs, providing specific billing, and so on. The use of CCSS No. 7 and the AMA record technique are discussed in more detail in Sections 3.4 and 3.6, respectively.

Standardized interfaces facilitate the use (purchase) of products already available on the market and are also a means of obtaining a good overall structure in the network. Sometimes, however, standardized protocols tend to be performance-consuming, regarding CPU capacity.

Being able to rely on a single vendor may make building an intelligent network platform and managing the network easier. Everything works more smoothly because you do not have too many different products or too many different vendors to cope with. It may not be good for your wallet in the short run, however, to be stuck with only one vendor.

The benefits of SPC exchanges, digital transmission, and PCM have become so obvious today that we often forget to mention them. It is true, however, that there are still many older non-SPC exchanges operating on networks, mainly as local exchanges, and their subscribers are often excluded from the use of many IN-based services due to older signaling protocols. These protocols make it impossible for a user to, for example, transfer the identity of the calling party out on the network, which is a requirement of many new services, both IN-based and node-based. Also, non-SPC exchanges often present users with restrictions to services that need high charging flexibility for different tariffs, like, for example, premium rate services.

Another problem with older exchanges is that they often have no capability for receiving DTMF pulses. A DTMF interface at the subscriber's phone or terminal is mandatory for using most value-added services (VASs), including those implemented on the intelligent network platform. While it is true that subscribers connected to nonSPC exchanges may use DTMF phones, signaling between the local and the transit exchange nonetheless presents a huge obstacle to using many services.

Calling line identification (CLI) is a very useful function in the implementation of VASs. This function makes it possible during call setup to have the calling party's identity available for use in different places or to have it transferred to the called party. This can be used in a variety of ways in combination with VASs and is further discussed in Section 6.2.2.

As was previously mentioned, it is not only the technical platform that is important in the implementation and marketing of IN-based services; a well-established organization that is already accustomed to conventional (not IN-based) services offers another excellent "platform" to start from.

Customers who have public switched telephone network (PSTN) access are undoubtedly the most interesting target group for IN-based services today and in the foreseeable future. This is because PSTN access (a normal telephone) can be found in practically every home and in every business today, while integrated services digital network (ISDN) access and ISDN terminals are not yet widespread and are not expected to become readily available for many years. To provide all customers with an ISDN access is expensive and will take a long time. However, IN-based services combined with access forms like ISDN, both narrowband and broadband, will probably become a greater market in the long run.

3.4 INTELLIGENT NETWORK IMPLEMENTATIONS WITH OR WITHOUT CCSS NO. 7

The most important items for standardization in intelligent networks are the interfaces, with the SSP/SCP interface being the most important and the most urgently needed in the short term. The reason for this is that it is the *basic* interface. If it is implemented, a first generation of intelligent network services, for example, translating services, like freephone and premium rate, can be opened. It is not necessary to have interfaces to, for example, service data points (SDP) and intelligent peripherals at once; these can be implemented in a second step.

3.4.1 Signaling Types

There are two basic parts of a communication link between two points on a network:

1. The speech channel, on which calls are conveyed (a "speech" channel is equivalent to other channels that transmit information between users, i.e., for data, text, pictures, and so on);
2. The signaling channel, which handles control information for the call, setting up a speech channel, releasing it, transferring numbers, and so on.

Signaling between exchanges can be carried out in two main ways, by using either

1. *Channel-associated signaling* (CAS), or
2. *Common channel signaling* (CCS).

CAS was formerly the only method for signaling offered on networks. It is called "channel-associated" because the signaling for the speech channel is conveyed in close relation to the speech channel itself. (In the old analog systems, signaling was carried out on the same physical wire the speech was transmitted on). CAS is also carried out in a fixed and static manner that does not allow, for example, new signals for new functions to be added. The number of possible messages that can be transferred is limited, and the capacity is low due to this rigid connection to only one speech channel. This is because CAS was established in the former analog networks in which there was no need to do more than transfer simple call setup information.

When digitization came into practice and the SPC technique was introduced, exchanges were interconnected via *pulse code modulation* (PCM) transmission techniques. Suddenly, in a 2-Mbps system, there was one 64-Kbps signaling channel for handling signaling for 30 speech channels. The first PCM systems were created according to the old CAS principle, but they eventually gave way to the CCS principle, which was more advantageous. CCS offers a more efficient use of signaling channels and the freedom to transfer call setup data and any other information via these channels. Moreover, as the SPC exchanges are controlled by CPUs, it is very beneficial to establish a direct data communication channel between the CPUs in both exchanges that need to communicate, i.e., to introduce CCS links.

A CCS system between two exchanges works with two autonomous signaling channels, one in each direction. These two channels are synchronized by the two nodes so that signals forwarded on one channel are acknowledged on the returning channel and vice versa. What is interesting to consider here is that if we only use the signaling channel to handle call setup data for the 30 speech channels, we normally only use 10–20% of the total capacity (which is 64 Kbps). The other 80–90% is free and can be used for other purposes, such as, for example, to transfer messages between SSPs and SCPs on an intelligent network.

The most widespread common channel signaling system in use today is CCSS No. 7, standardized by ITU-T. It is widespread in many networks, but, nevertheless, there are a lot of networks that still use older signaling systems. The advantages of using CCSS No. 7 are obvious, and they are not only related to the building of intelligent networks. All networks are more or less heading towards CCSS No. 7.

3.4.2 Common Channel Signaling System No. 7

This section does not attempt to provide an in depth description of CCSS No. 7, which would be a very complex task. However, as CCSS No. 7 is very important in intelligent network implementations, we examine the parts that are essential to implementation. Readers who are interested in acquiring a more comprehensive knowledge of CCSS No. 7, should refer to [1–3].

In addition to aiding intelligent network implementation, CCSS No. 7 also offers the freedom of transferring signaling between the two exchanges in three different modes (see Figure 3.1):

1. An associated mode (path "a" in Figure 3.1);
2. A quasiassociated mode (a predetermined path "b" in Figure 3.1);
3. A nonassociated mode (path "b" chosen in some more dynamic way, maybe randomly).

Do not confuse the signaling modes with the CAS, which, as we saw in Section 3.4.1, is something entirely different.

In the associated mode, signaling *always* takes the same path (not the same channel) as the speech channel on the network. In the quasiassociated and nonassociated modes, signaling is routed independently of the speech channel, that is, speech can go one way and signaling another. In the quasiassociated mode, there are predetermined paths for signaling to follow. The nonassociated mode points to a more general situation in the future in which the goal is to let the signaling setup be completely independent and work autonomously. Quasiassociated could be described as a special case of, or a first step towards, the nonassociated mode.

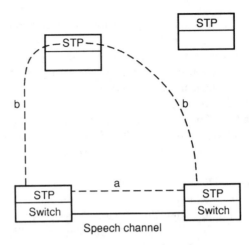

Speech channel

STP = Signaling transfer point

Figure 3.1 Signaling modes in CCSS No. 7.

The quasiassociated and later nonassociated modes separate signaling paths from the speech paths. Therefore, it is very convenient to consider two different (functional) networks, a transmission network for transferring speech and data and a signaling network. The nodes in the network that handle (originate and terminate) signaling in a CCSS No. 7 network are called SPs (signaling points). To transfer the signaling between the SPs, STPs (signaling transfer points) may be used, but not necessarily. The STPs then form, together with the SPs, a separate network for signaling, for our purposes, called a signaling network. An SP, or an STP, is often a normal exchange as a transit exchange, but it can also be a node of its own.

For the purposes of this text, we will regard the signaling network as a network of its own—a 64-Kbps data communication network capable of handling, for example, the communication between the SSP and the SCP.

Figure 3.2 illustrates the structure of CCSS No. 7. The part that is of primary interest in relation to intelligent networks is the *IN application protocol* (INAP) within the component sublayer of the *transaction capabilities application part (*TCAP) on the application layer, that is, layer 7 of the OSI model. The figure does not show every detail of CCSS No. 7 structure; rather, it indicates where the parts that are relevant to intelligent networks are located with respect to the OSI 7-layer model and the signaling 4-layer model. The acronyms in the figure stand for:

Acronym	Term
ASE	Application service element
TCAP	Transaction capability application part
ISP	Intermediate service part

ISUP	ISDN service user part
SCCP	Signaling connection and control part
TUP	Telephone user part
DUP	Data user part
MTP	Message transfer part

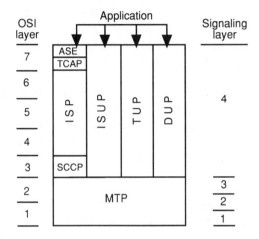

Figure 3.2 The structure of CCSS No. 7.

The *telephone user part* (TUP) is employed by existing networks that use CCSS No. 7 for normal speech traffic (normal calls). Networks can also use a modified TUP for communication between the SSP and the SCP. This is described in more detail later in the text.

3.4.3 Implementations that Use Service Control Points

When SCPs are used to control the intelligence for a service, only signaling is used between the SSP and the SCP. This section describes three approaches to implementing intelligent networks with SCPs; the differences between the three approaches depend on the signaling capabilities available. To facilitate understanding, the examples describe the implementation of a service that uses number-translating, the freephone service. The same approaches are also valid for other number-translating services, such as premium rate and universal number.

A Freephone Implementation that Uses CCSS No. 7 with the TCAP and the INAP

Using TCAP/INAP offers numerous benefits. First, it provides a standardized protocol, which allows an open approach to different systems and different vendors of intelligent network equipment. Second, and more important, it offers the capability of transferring other types of information (not circuit-related information) besides information related to call setup in the interface. Third, it provides more security in the communication from the protocols. Fourth, services that use devices to broadcast recorded announcements to subscribers during call setup (examples are provided in Chapter 4), can use the signaling network to control the announcement device in the subscriber's own transit or local exchange by a remote SCP.

Figure 3.3 shows a typical freephone service setup that uses the TCAP/INAP on a CCSS No. 7 network. This solution allows signaling and speech to be separated; the TCAP/INAP works as a protocol conveyed by signaling channels, which, in turn, are regarded as a *separate network* (from the speech network), or a signaling network. The freephone service steps, illustrated in Figure 3.3, are:

1. The local exchange hosting the calling party recognizes a freephone call from subscriber A by analyzing the called number.
2. When it finds a freephone number, the local exchange routes the call (both speech and signaling) to an SSP, which is often the transit exchange.
3. At the SSP, call setup is suspended while a setup to the SCP is made via the CCSS No. 7 signaling network, using the messages conveyed by the TCAP/INAP.
4. In the SCP, the freephone number is translated to the C number (i.e., the destination number for the freephone call).
5. The C number is returned on the signaling network to the SSP that required the translation.
6. When the SSP receives the C number back, normal call setup (signaling and speech) is resumed to connect to the destination local exchange and, finally, the C subscriber.

The TCAP/INAP is the recommended way of signaling between SSPs and SCPs and, in the future, it will undoubtedly replace all other methods. There have not been a lot of TCAP/INAP implementations to learn from to date (December 1993), however, and it is to be expected that they will use more loads in the CPUs in the nodes than the modified TUP (see below) and similar implementations. This is most important, of course, in implementations of mass calling services, like televoting, which can entail sudden bursts of traffic. Perhaps the best solution is to have two protocols on the intelligent network, one that guarantees high security and performance of features (TCAP/INAP) and one that provides less security but offers fast response when that is the essential requirement. This is also discussed in Section 7.4.1.

A Freephone Implementation that Uses CCSS No. 7 and a Modified TUP

The normal procedure when the CCSS No. 7 has a TUP but not a TCAP/INAP is to add some extra signals to the TUP exclusively for handling the communication between the SSP and the SCP. A freephone number setup with a TUP modified in this way follows exactly the sequence illustrated in Figure 3.3. The only difference is the method of signaling

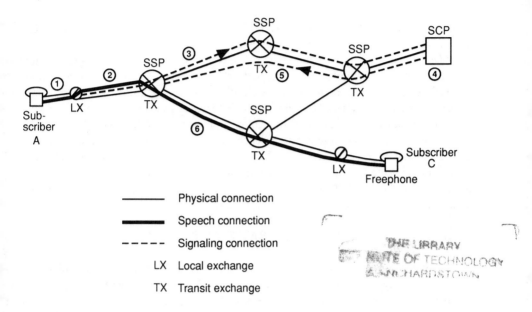

Physical connection

Speech connection

----- Signaling connection

LX Local exchange

TX Transit exchange

Figure 3.3 A freephone setup with a CCSS No. 7 network and an SCP solution.

between the SSP and the SCP. A modified TUP solution uses the facility within CCSS No. 7 that is offered by the *quasiassociated or nonassociated modes* (see Section 3.4.1 and Figure 3.1) to create different paths for speech and signaling. Only the signaling channel is required, however, to transfer a number to the SCP for translation into another number, which must be returned to indicate where the freephone call must be routed.

Since the speech channel is not needed and the distance between the calling SSP and the called SCP could be very large, it is best simply to have no speech channel at all and to use only signaling. But this is not easily done in networks in which CCSS No. 7 was mainly implemented to set up normal calls. Signaling requires a physical connection. There is a method, however, that can solve this. (It was used in Sweden and some other networks in the first years of intelligent networks.)

The software in SSPs and in the SCP defines direct routes between all SSPs and the SCP. In practice, however, there are no physical connections, only hardware devices on both sides that are physically loop-connected back to the SSP/SCP. In other words, the SSPs and the SCP "believe" there are direct speech channels between them, but there are only loop-connected devices at both ends (see Figure 3.3). When a number translation is needed, the SSP at the calling party's end calls the SCP, just as if it were a normal telephone call, and a "speech channel" is reserved at both ends of the SSP/SCP connection (but no transmission link exists). As the speech channel (which does not physically exist) is reserved at both ends, signaling can be exchanged between SSP and SCP (dotted lines labeled 3, 4, and 5 in Figure 3.3). CCSS No 7 uses the quasiassociated or nonassociated mode. When the number is returned to the SSP, the freephone call is set up, just like any normal call setup to C (the destination number).

The advantages of this solution are that it is easy to implement and takes less time to start up with (a simple) intelligent network, compared to the implementation of TCAP/INAP (of course, you must have a CCSS No. 7 network). It is usually not very per-formance-consuming in the CPUs and it provides short answering times from the SCP.

Drawbacks to the modified TUP solution, compared to the TCAP/INAP solution, are that the type of information that may be exchanged between the SSP and the SCP is very limited and that, after delivering a translated number, the SCP loses control over the rest of the service. This means, for example, that the SCP has no knowledge of the remaining part of the setup of the freephone number. It does not know if there was any answer, if the call failed because of congestion or another reason, and so on.

A Freephone Implementation when CCSS No. 7 Is Not Available

If a freephone implementation employs the SCP, but CCSS No. 7 is not available, you must create a way of communicating between the SSP and the SCP without a speech chan-nel, that is, by using a communication link outside the network. (See Figure 3.4.) Usually a data communication X.25 network can be used (X.25 is described in many books, for in-stance [2]). The disadvantage of this method is somewhat longer response times compared to the CCSS No. 7 solutions. However, if there are no alternatives and if the response times are acceptable—for example, if no real-time requirements exist and a delay of 3–5 seconds is acceptable, or if the communication is a facsimile or a file transfer—this is a possible interim solution. Many networks today have an X.25 network that interconnects the switches and the SCPs. The X.25 network is usually used for network management and exchange operations. It should be easy to use as an interim solution for communication be-tween SSPs and SCPs before CCSS No. 7 is available. Remember too that there are new players in the intelligent network market. Data communication companies are starting to develop SCPs, and they have experience and a long tradition of protocols like X.25.

With this method, it is quite simple to set up a freephone call:

1. The local exchange hosting the calling party recognizes a freephone call from subscriber A by analyzing the called number.
2. When it finds a freephone number, the local exchange routes the call (both speech and signaling) to an SSP, which is often the transit exchange.
3. At the SSP, call setup is suspended, while a setup through a separated data communication network (preferably an X.25) is made to the SCP.
4. At the SCP, the freephone number is translated to C (the destination number).
5. The C number is returned on the data communication network to the SSP that requested the translation.
6. When the SSP gets the C number back, normal call setup (signaling and speech) is resumed to connect to the destination local exchange and, finally, the C number subscriber.

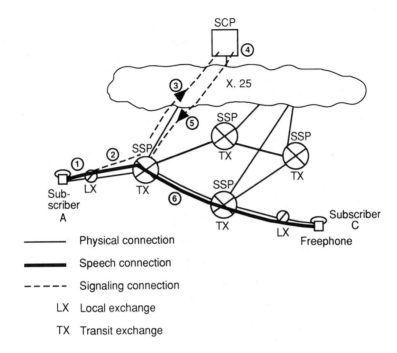

Figure 3.4 Physical connection, Speech connection, Signaling connection legend

- ——— Physical connection
- ■■■■ Speech connection
- – – – – Signaling connection
- LX Local exchange
- TX Transit exchange

Figure 3.4 A freephone setup using the SCP concept, when CCSS No. 7 is not available.

3.4.4 Implementations that Use Service Switching and Control Points

A service switching and control point (SSCP) is a physical node that contains both the service switching function (SSF) and the service control function (SCF). The interface between the two is an internal software interface within the node itself.

SSCPs are often implemented when no CCSS No. 7 exists on the network, because it is convenient to route the speech channel to the node where SCF is located. SSCPs are also the oldest way of implementing intelligent networks; even before the intelligent network concept came into widespread use, the SSCP was a dedicated service node in a central location on the network.

In addition, if a service requires an announcement device, a single announcement device situated at the SSCP can serve a large area (see also Section 7.5.1). Traffic for the service, including signal flow on the speech channel, would be routed to the host SSCP of the device.

In implementations that employ the SSCP solution (see Figure 3.5), the freephone call service works as follows:

1. The local exchange hosting the calling party recognizes a freephone call from subscriber A by analyzing the called number.
2. When it finds a freephone number, the local exchange routes the call (both speech and signaling) to the transit exchange.

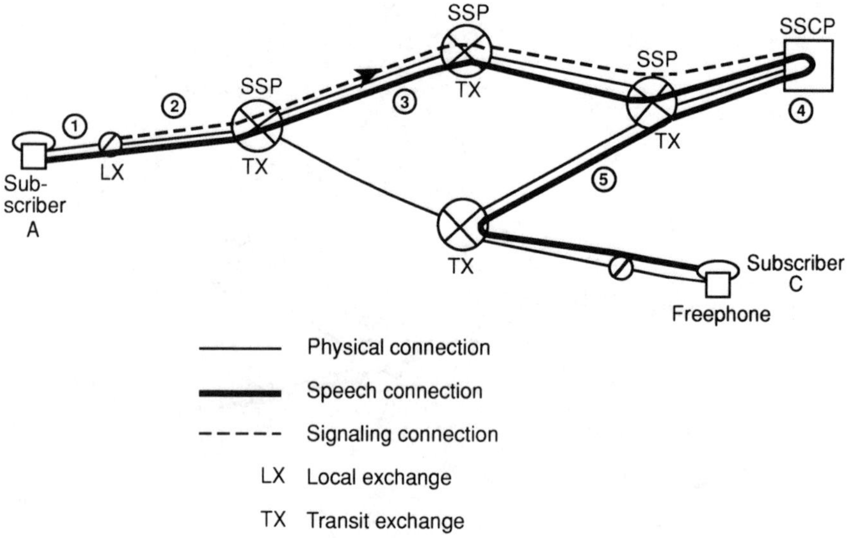

Figure 3.5 A freephone setup using the SSCP solution.

3. At the transit exchange, a normal call through the network is set up to the SSF in the SSCP.
4. In the SSCP, the SSF calls the SCF to get the freephone number translated to C (the destination number). After that, the C number is returned to the SSF (all of these are examples of internal signaling within the SSCP).
5. When the SSF gets the C number back, a normal call (signaling and speech) is set up from the SSCP to the C number subscriber.

It can be seen that the speech path must follow signaling exactly, that is, speech and signaling cannot be separated from each other with the SSCP method. This means that the speech path has to be routed together with the signaling path to the SSCP (where the SSF and the SCF are). In other words, the speech path has to be set up through the SSCP before it connects to the C number subscriber, which often is a waste of speech paths, because there is the risk that the total way in the network for the speech will be very long, compared to the SCP solution. This can cause problems, as speech channels in the network may run out much easier in this way. Consequently, this can cause congestion for other call setups.

3.4.5 Mated Pairs of Service Control Points and Service Switching and Control Points

When an SCP or an SSCP is introduced for a certain service, it is often given the responsibility of all handling and control of the service, for the entire network. This is a classic case of "putting all of our eggs in one basket." If the SCP/SSCP fails, availability of the service is lost throughout the *entire* network. To avoid this, some kind of redundancy is necessary. A very handy way of accomplishing this is to work with *mated pairs of SCPs/SSCPs*. Two SCPs/SSCPs working as a mated pair should:

- Be physically separated on the network, and
- At every moment, possess identical software and subscriber data.

Distribution of service traffic differs, though, in the SCP and the SSCP solutions. For mated SCPs, the solution to calling the SCP is often very simple because the normal routing schemes on the network can be used. In the SSPs, the two SCPs in a mated pair can be defined as *first* and *second* choice due to routing algorithms. When the SSP is calling an SCP, the first attempt tries to reach the first choice. If for some reason it does not answer—due, perhaps, to failure or congestion—the network automatically tries the second choice.

With mated SSCPs, however, it is not always as straightforward. In this case, the SSF is physically located with the SCF. The solution greatly depends on the signaling used. What is important, in the case of a failure in the SSCP, is whether we notice it in the calling node and can reroute the call to the other SSCP, as in the SCP case, or notice the failure first when reaching the SSCP itself. The latter case is more complicated, as several solutions exist.

Customer control, management functions, service setup using mated pairs, and possible implementation solutions of the future are further discussed in Chapter 7, Sections 7.3.2 and 7.4.2.

3.5 TELEPHONE TRAFFIC THEORY APPLIED TO NEW SERVICES

Until recently, telephone traffic theory, for example, calculating the number of channels in routes between telephone exchanges, dimensioning exchanges themselves, or calculating congestion, was relatively straightforward. We were able to predict traffic relatively well by using general telephone traffic theory. Experience showed that traffic peaks during certain hours of the day, which has led to the "busy hour" concept—the hour of the day with the highest traffic. The duration of calls has always been assumed to average 2–3 minutes, which is true for a *basic call* [3].

The busy hour traffic concept has been used to calculate the load on a network, to allow dimensioning of all transmission and switching equipment. Basic telephone calls have been the dominating traffic on the networks. When value-added services were introduced on networks, such as call forwarding, call transfer, call completion to busy subscribers, and other node-based services for which control is located in the local exchange, traffic theory was still valid. This was because these services were closely *related to basic calls*, that is, the services were initiated by basic calls and they followed the *same load pattern*. Furthermore, the network load from value-added services has been rather limited compared to the load from basic calls.

Now, however, a new type of service has been introduced that is independent and not directly associated with basic calls. Televoting and premium rate are good examples of these new services. And these services, combined with a higher relative load on the networks, results in *a completely new load pattern* on the network. The traffic profile of these new services is much more hazardous to the network than basic-call traffic. The new services often result in traffic bursts, and it is not always possible to predict exactly when a traffic burst will occur.

As an example, to calculate the number of calls per second, which is very important in determining the CPU (processor) load on telephone exchanges or in SCPs, the following very simple formula is used:

$$S = N * \frac{a}{T}$$

N is the number of subscribers; a is the average traffic generated per subscriber, expressed in erlang; and T is the average time duration for a call. S is the value we are seeking, the number of calls per second, which constitutes the load on the processor in the local exchange.

Let us consider the following very normal values in a local exchange handling basic telephony:

N = 50,000 subscribers
a = 0.1 erlang per subscriber during the *busy hour*
T = 120 seconds

This gives us a value for S of 42 calls per second.

Let us assume that, in dimensioning the local exchange, we want a processor that is capable of handling perhaps 150 calls per second, which should provide a sufficient margin. But will it? Let us see what happens when we introduce televoting and run it on a Saturday evening during a popular television program. Votes can be collected in many places on the network—at the local exchange itself, at the SCP, and so on—but all "calls" (votes) from subscribers must pass through the local exchange.

It is 8 pm on Saturday evening and a very popular television program is being broadcast live. At 8:15 pm, the host begins a game by *showing a televoting number on the television screen* and telling the audience they can win $100,000 by calling the number immediately. Only the first 100 people that reach the television program will win, however. Let us assume that *20% of the subscribers* are watching the program and want to win the $100,000. The rest are not watching the program or are not interested in playing the game. Accordingly, we will have two groups of subscribers with two types of behavior: those who call the televoting number and those who do not, but want to continue their normal communications.

If 20% of the subscribers call the televoting number, 10,000 calls will have to be routed through the local exchange. With all these calls coming at the same time, congestion is almost inevitable. If you encounter congestion in your attempt to win $100,000, you will probably continue to call again and again for a number of minutes. People trying to get through will probably make a call attempt as often as every ten seconds. This will last for at least a couple of minutes, until they reach the host or give up. Now, while ten seconds between each call attempt may seem to be a short interval, we must remember that *the smart callers* probably entered the number as an *abbreviated number on their phone* in order to call more quickly.

This gives us a situation in which 10,000 subscribers are making call attempts for two minutes, over and over again, every ten seconds, which corresponds to a load of about *1,000 calls per second* on the CPU in our local exchange! (Remember that the CPU in the exchange was dimensioned for 150 calls per second.) To this load must be added the normal traffic from the 40,000 subscribers who are not calling the televoting number. While this traffic will be negligible compared to the 1,000 televoting calls, it is clear that no normal traffic will be able to get through the local exchange during the televoting session. The overload could be very serious, especially in the case of emergency calls to the fire department, a doctor, or the police.

This example illustrates why network traffic must be limited, in one way or another, to avoid reaching full capacity. On an intelligent network, this can be done in many different ways:

- By introducing *call gapping*, that is, by slowing down the traffic between the SSP and the SCP and not letting it run the risk of an overload. Call gapping involves spreading the traffic in time by buffering messages until traffic is calmer.
- By introducing a *limiter in the CPUs,* in the switches and in the SCPs, which, for example, only allows the CPU load from a certain service to reach 50% of its total capacity. The rest is left for other services or normal calls.
- By introducing *priority levels* in the network (in SCPs, SSPs, and local exchanges), whereby certain number series or certain numbers have higher priority than others.

Any combination of these methods can also be used. Chapter 7 further discusses controlling network traffic.

Another conclusion that can be drawn from the televoting example is that serious consideration must be given to how future networks are dimensioned. We need a *new view of the telephone traffic theory*, including new tools for dimensioning the different components in networks. We need to consider shorter time intervals, as traffic can change more rapidly. In fact, traffic dimensioning in the future will probably demand development of a "busy minute" concept to correspond to today's busy hour concept.

Televoting is not the only service that creates a new type of traffic pattern on the networks. An increase in other types of services, including wider use of other mass calling services, like premium rate, and increasingly greater data communication traffic also create new traffic patterns.

3.6 INTELLIGENT NETWORK MANAGEMENT FUNCTIONS

3.6.1 The Service Management Point

The *service management point* (SMP), hosting the *service management function* (SMF), is what we could call "the spider in the network." Although there will be numerous SSPs and quite a few SCPs in networks in the future, there will probably be only one SMP (one SMF) (see Section 2.2.1 and Figure 2.2). The SMP is where the total overview of an intelligent network can be obtained.

The main task of the SMP is to deploy and manage IN-based services. The SMP maintains control of the whole intelligent network system. Typical functions it performs include:

- Distributes created service logic to SCPs;
- Maintains a copy of the network configuration of the IN-based services, so it can update SCPs in case of failures;
- Manages control data generated by network operation staff and customers via the customer control function;
- Removes service logic from SCPs;
- Controls which services occupy which SCPs;
- Handles mated pairs of SCPs (simultaneous updating, and so on).

The SMP is discussed further in Chapter 7.

3.6.2 Charging, Billing, and Administration

As the discussion in Chapter 2 points out, in order to ensure rapid implementation, change, and withdrawal of services, the whole chain of network systems must follow the intelligent network approach. Of all the systems involved in a total intelligent network solution, the charging and billing systems are often the oldest. This is especially true in Europe. In the United States, the AMA record technique is widespread, but in Europe most networks still use the old call metering technique for charging. (Charging and billing, including discussions of the future, are also described in Section 6.3.3.)

The Call Metering Technique

The call metering technique for charging a call or a service is highly inflexible and static. A call meter is set up for every subscriber at the local exchange, which counts the pulses generated for the activities carried out. Certain rules determine how many and how often pulses are generated by different types of traffic, such as local calls, transit calls, international calls, or network services. The call meter is read periodically, the difference from the last reading is calculated, and a bill is sent to the subscriber based on the number of pulses generated for the period. The reading cannot indicate the type of call or service that generated the pulses, nor can it generate, for example, call specification information. The call metering technique can be regarded as an *online charging method*, that is, a method whereby users are charged immediately.

Other charging issues include traffic that moves across networks and service providers who offer premium rate services with shared revenues. In these instances, money must be transferred from the party charging the subscriber to the other parties involved in the call setup or the service. The amount of money to be transferred must be decided in some way. This may be done by setting up two call meters that are stepped at different intervals. One would indicate what the subscriber should pay and the other would show how much should be transferred to the other network operator or service provider.

The Automatic Message Accounting Record Technique

If the call metering technique can be described as an online charging method, the automatic message accounting (AMA) record technique (also called toll ticketing) is an *offline charging method*. The basic reason for introducing the AMA record technique is to move all charging off the network and onto an offline system. By this method, the network would only provide the *raw data* about every activity a user completes, such as:

- The type of activity (for example, a normal call or a service) and also the call attempts that fail;
- Calling party (subscriber A);
- Called party (subscriber B);
- Subscribers C and D (that is, other parties involved), if relevant;
- The date and time of the activity;
- The duration of the activity.

This information is transferred to the offline system for charging and billing. The transfer is carried out via a data communication channel, which can reside on a separate data communication network or, in future, on the data communication network provided by the CCSS No. 7 signaling system (described in Section 3.4).

The benefits of using the AMA record technique are obvious. One benefit is the ability to adapt the charging and billing system to new services quickly. Other benefits include:

- Different prices can be set for different customers. For example, discounts may be granted to major customers.
- Parties can be charged separately or the bill can be split between parties.
- The charging party can be changed during a call.
- A third party can be charged.
- Failed calls can be recorded.
- Call statistics can be kept, if the data above is recorded.
- Customers can have all their calls and activities specified on their bills.

A general conclusion can be made about charging and billing systems. To introduce a new service on the intelligent network platform quickly requires the same degree of flexibility required by the rapid updating of charging and billing systems. Consequently, the AMA record technique is a very good candidate for both national and international networks. (The AMA record technique and future evolution needs are further discussed in Section 6.3.3.)

REFERENCES

[1]   Manterfield, Richard J., *Common Channel Signaling*, IEE Telecommunication Series 26, Peter Peregrinus Ltd., London, United Kingdom.

[2]   Freeman, Roger L., *Telecommunication System Engineering*, Wiley Series in Communications, John Wiley & Sons Ltd., New York, 1989.

[3]   Clark, Martin P., *Networks and Telecommunications: Design and Operation*, John Wiley & Sons Ltd., Chichester, 1991.

Chapter 4

Exploitation of Intelligent Networks in Services Today

This chapter describes the current situation (autumn 1993) regarding the introduction of IN-based services, including some of the most common services.

The *first common* set of services to be generally implemented in many networks will most likely be Capability Set 1 (CS1), created by ITU-T and ETSI. CS1 will not become a standard, it will instead be a set of services and SIBs for *benchmark testing*. If the services in CS1 can be built on a network, the network conforms to CS1. Instead, what will be standardized are the physical interfaces, for example, the Intelligent Network Application Protocol (INAP) within the Transaction Capability Application Part (TCAP) in OSI layer 7. Because CS1 was not stable until 1993, it will take a few years to implement it on the networks. (CS1 services, features, and SIBs are described in Section 2.4.4 and in the Appendix.)

The *reality* in networks today (1993) is quite the opposite; as CS1 is stable first now, services, features, and service-independent building blocks (SIBs) are all vendor-dependent, that is, they are dependent upon the way vendors implemented the IN-based services. There are different sets of services on networks, and the same service can be implemented by different methods—IN-based, node-based, or terminal-based—on different networks. Moreover, in the case of IN-based services, the different implementations share no commonality of SIBs. Not even the level (the size) or the functionality of the SIBs are the same. Some SIBs are small (only a few lines of software code); some are larger and can almost be regarded as features.

This chapter focuses on a pragmatic situation involving the most common services implemented today or that will be implemented in the near future. Both services in general and services based on CS1 are described. Because the IN-based services will exist in an environment that includes other (older) services, a total service approach is taken. Consequently, services outside the IN platform are also handled.

4.1 SERVICES AVAILABLE BEFORE INTELLIGENT NETWORKS

4.1.1 General Aspects

Before describing how to best exploit intelligent network platforms to build services, we should focus briefly on the conventional services that existed prior to the introduction of intelligent networks.

To review, Section 2.2 described the three different methods for implementing services: (1) terminal-based, (2) node-based, and (3) IN-based (also called network-based).

Conventional, non-IN-based services were introduced in the 70s and 80s, usually, first as node-based services. Later, some of these services, such as abbreviated dialing and repetition of last dialed number, were moved to terminals, and they became terminal-based services. These conventional services will not be removed from the network simply because intelligent networks are introduced. In future, of course, some of them may *disappear,* because they no longer fulfill modern requirements, or because their functions are extended, or because they are being replaced by other new services. The new services could, in turn, be IN-based, terminal-based, or node-based.

So, instead of being discarded, some services implemented on the network before the introduction of IN, mainly node-based services, will probably be replaced by services implemented either on the terminals or on the intelligent network platform. The advantage of moving services to the terminals is that it minimizes signaling needs between the terminal and the local exchange, for example, when setting up or changing a list of abbreviated numbers. Porting of services to the intelligent network platform will allow a network operator to obtain a better overview and provide improved service management. There will also be a greater demand for interoperability of new services and the older, node-based services.

To sum up, the general trend today is to abandon the old *one and only* way of implementing services on nodes and to consider, when justified, terminal-based or IN-based implementation of conventional services (see Figure 4.1). This trend is further discussed in Chapter 6.

The service repertoire built *before intelligent network technology* may vary considerably from network to network. Some services, however, are or will be present in almost all networks. Without claiming to be complete, this section describes a majority of these services. The function of a basic service may also have small variations from network to network *from the user's point of view.*

In IN-based implementation there is *a planned way* of allowing variations in a particular service, to fulfill different customers' needs. This is one of the *benefits* of building services with the intelligent network platform, namely, the concept of *service provisioning and customization.* Basic services are tailored to suit the special requirements of individual customers. This concept is discussed in Section 4.2, and examples are provided in Section 4.3.

But even two services that appear identical to users might, because of different implementation methods, behave differently, for example, in the way they interact with other services. These differences, which are *implementation-dependent,* are referred to in this

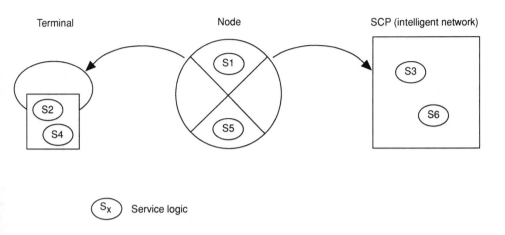

Figure 4.1 Traditional, node-based implementation is no longer the only way today.

text as *service dialects*. Service dialects existed long before intelligent networks were introduced. And, with intelligent networks, service dialects will continue to exist, because many different intelligent network platforms will exist. (Customized services and service dialects are also discussed in Section 6.4, which deals with service interworking.)

4.1.2 Conventional Non-IN-Based Services

While conventional services can be either node-based or terminal-based, the majority today is node-based. Figure 4.2 shows from above the user interface (from the service user's point of view) and from below how deeply the service is rooted (integrated) in the network. "Deeply rooted" is an apt description because it refers to the fact that once a service is embedded (rooted) in the network, it is very difficult to change it. Originally, as we discussed above, the trend was to implement all services on the nodes. However, extensive signaling between nodes and terminals will result if the services become popular. In this case, it is more convenient to move whatever can be moved from the node to the terminal to decrease the load on the nodes and signaling systems. That is why some services today are only implemented in terminals. Good examples of services that are commonly shifted from node-based to terminal-based implementation today include repetition of last dialed number and abbreviated dialing. Figure 4.3 illustrates this situation. Once terminal-based services have been implemented, they are also not easily changed due to, among other things, the large number of terminals spread among subscribers.

The following list describes conventional services that were common in most networks prior to intelligent networks. When possible, the most common method of implementation is mentioned as well. To simplify matters, "A" represents the calling party, "B" represents the called party, and "C" represents the next party involved, where applicable.

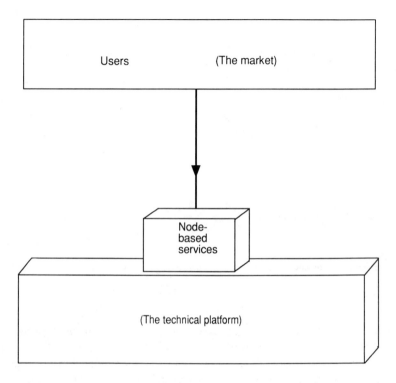

Figure 4.2 Node-based implementation of services.

- *Abbreviated dialing (short number).* Telephone numbers that A dials frequently may be keyed in as short codes (only 3–4 digits instead of the 10–20 that normally need to be dialed). The short code is then translated to the actual number. Although this service was originally implemented in the local exchange, it is an excellent example of a service that ought to be (and today often is) implemented in the terminal (the telephone) instead.
- *Automatic alarm call (wake-up service).* This service was first known as the wake-up service. Today, however, it is more generally used to activate a terminal at a predetermined time. At the indicated time, the service (network) initiates a call to the terminal.

  It is important to restrict this service from being combined with services like *call forwarding*. If those two services can be used together, a practical joker can order a wake-up call for another person. To safeguard against inappropriate use of the service, users can only order the service from the phone that is to be called (alarmed). (See also Section 6.4, which deals with service interaction problems.)

  Today, this service is usually implemented in the local exchange, but new terminals could easily include it in future. Implementing the alarm function in terminals is technically very easy and can decrease the need for signaling in the networks.

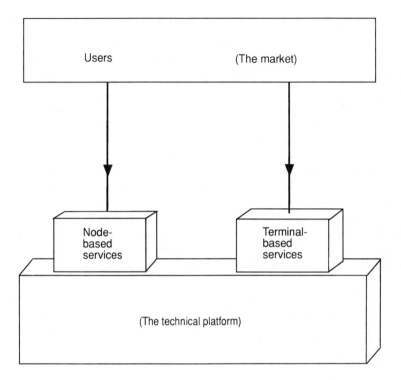

Figure 4.3 Node-based and terminal-based implementation of services.

- *Call completion to busy subscriber* (CCBS). If A calls B and B is busy, A may initiate *call completion to busy subscriber* from her terminal before putting down her handset. When B becomes free, A's telephone rings in a special manner (a distinctive tone). If A still wants to reach B, she has to lift her handset within a certain period of time to be connected to B. Once B lifts his handset a call is established.

    Today, this service is easy to implement if A and B are located at the same local exchange. It is more problematic, however, if two separate local exchanges are involved, as the means of transfer becomes more complicated. From the implementation point of view, the service is divided between the local exchanges of A and B.

- *Call completion when no reply* (CCNR). If A calls B and B does not answer, A may initiate *call completion when no reply* from her terminal before putting down her handset. Now the network must watch to see when B becomes available. It does this by noticing when B lifts off and releases his handset. Once this happens, A's telephone rings in a special manner (a distinctive tone). If A still wants to reach B, she has to lift her handset within a certain period of time to be connected to B. Once B lifts his handset a call is established.

This service is easy to implement if A and B are at the same local exchange. It is more problematic, however, if two separate local exchanges are involved because the signals between exchanges are more complicated. From the implementation point of view, this service is divided between the local exchanges of A and B.

- *Call forwarding unconditional* (CFU). If *call forwarding unconditional* is initiated, all incoming calls to B will be forwarded unconditionally to C. The C number must be initiated (and removed) from B's telephone. Today, this service is often implemented at B's local exchange.

  A service with the same functionality can easily be implemented on the intelligent network platform by using number translation features. Extra features, like remote control via the customer control interface may then be easily added, if desired.

- *Call forwarding (with condition)* (CF). Incoming calls to B will, if the necessary conditions are fulfilled, be forwarded to C. The C number must be initiated (and removed) from B's telephone. The most common conditions that require call forwarding today are if B (1) is busy or (2) does not answer within a certain period of time or after a certain number of signals (ringing cycles). This service is often implemented at B's local exchange today.

- *Call hold* (CH). A party uses the call hold service to temporarily interrupt one call in order to initiate a second call or service. CH is often used in combination with the *multiparty* or *call transfer* services described below. It often has a node-based implementation today.

- *Call pickup.* This service makes it possible for B to receive an incoming call on any phone. Of course, this assumes that B is aware that he is being called. Call pickup can be used, for instance, in a large office where B can hear his phone ring but is too far away to reach it. He can receive the call by picking up the phone nearest him and keying in a code (password). This service has often a node-based implementation today.

  This service can also be combined with a *paging system*. When B hears the tone on his pager, he can pick up any phone, dial a code (password), and the incoming call will be connected to that phone.

- *Call screening services (call barring).* This term refers to a number of services with the same goals: to eliminate A's ability to make dedicated outgoing calls or to prevent B from receiving dedicated incoming calls. For example, A might be forbidden to make international calls because they are very expensive. *Outgoing call screening* services are often used when users have access to phones but do not have to pay for calls, for example, employees. *Incoming call screening* is often used to protect someone from receiving calls from certain A numbers.

  Outgoing call screening can be:

  □ Geographic, where A is only allowed to call within a certain area.

  □ Limited in number, where A is only allowed to call a predetermined group of subscribers. This service is also called *closed user group* (CUG).

  □ Dependent on duration, where A is only allowed to speak for a certain period of time before the call is interrupted.

□ Dependent on time, where A is only allowed to make calls within certain time periods.

□ Restricted, where A is not allowed to call expensive services, such as premium rate services.

Incoming call screening can be:

□ Geographic, where B only wants to receive calls from a certain area or areas.

□ Personal, where B wants to or does not want to receive calls from certain A subscribers.

These services often have a node-based implementation today, that is, in the local nodes of A or B.

- *Call transfer* (CT). Call transfer, in contrast to call forwarding, is always carried out during an ongoing call. It is commonly used with call hold (see description). A calls B, and during the call B initiates a call hold for a moment (temporarily leaving contact with A) to make a call to C, perhaps to ask C a question. B now can *pendulate* between A and C. If B decides to connect A directly to C, B can initiate a call transfer to C, that is, B leaves the call and A and C are connected.

  This service is often implemented at the local exchange level today.

- *Call waiting* (CW). A makes a call to B, who is already involved in a telephone call with C. If B subscribes to the *call waiting* service, she is notified (very discretely) that someone is trying to call her during her conversation with C. At the same time, A will not meet a busy tone, just a normal ringing tone. In this situation, B can either ignore the new call or ask C to hold for a moment while she switches to the new call. After talking to A, B can return to her original call with C.

  This service often has a node-based implementation today.

- *Fixed destination call (hot line).* This service makes it possible for A to initiate a call to a predetermined B subscriber without dialing a number. It can be done at the instant that A lifts her handset or after a few seconds, to permit A to make a normal call. This service often has a node-based implementation in the local node of A.

- *Immediate service (ringback price).* This service allows a caller to obtain information about the price of a call as soon as it is finished, for example, when the caller is using someone else's phone. The service often has a node-based implementation.

- *Malicious call identification.* This service makes it possible for the B subscriber to request a trace of a malicious call. It is usually used when a subscriber receives obscene calls or the like. After receiving an obscene call (A probably puts down the handset after a short period), B has a certain period of time to initiate the service from the terminal. Once this has been done, the A number will be registered by the network and made available to B later on. This service has often a node-based implementation today.

- *Multiparty service.* This service involves calls in which more than two parties are involved, usually *threeparty calls*. Actually, the service performs more than one function. During a normal two-party call from A to B, it allows one party to initiate a call hold (temporarily interrupt the call, see description) and make an inquiry to a third party (C). She can then alternate between the two calls, establish a three-party call, or leave the call and allow the other two to continue (call transfer). This service often has a node-based implementation today.

- *Repetition of last dialed number (redial facility).* When A calls B and does not reach B for some reason (B is busy or not at home) or if A wants to make a second call to B, this service comes in very handy. It simply remembers the last number A dialed. By dialing a special code at her terminal, A can initiate a new call to B automatically.

    This service was initially implemented in the local exchange; however, it is an excellent example of a service that should be (and today usually is) implemented in the terminal (the telephone) instead.

- *Remote control facility.* This mechanism allows a remote capability to be established for many of the services described in this list, even without implementation of an intelligent network and service control points with customer control facilities. Normally, to initiate any of these services, the user must be physically present at the right phone. With a remote control facility, however, the subscriber may initiate a service from any phone. This capability can be implemented in many ways. It has been available in Sweden since 1992 as *remote control of call forwarding unconditional.*

The technical solution, which is applicable to other services as well, is accomplished in the following manner. If, for example, a call forwarding unconditional should be done on someone's home phone while he is at another location, he may reach the remote control facility by making a freephone call. This freephone call is set up to separate control equipment on the network. Such a call may be made from any DTMF phone. After providing a password, for identification and validation, the subscriber may initiate call forwarding to any other telephone number in the world. The service is initiated by the control equipment, which creates the same sequence of star (*), pound sign (#), and the digits that the subscriber would have inserted if at home, and sends the sequence to the subscriber's local exchange.

Of course, the remote control facility is very easy to implement when customer control on an intelligent network is used. Services that are similarly complemented by remote control, for example, call forwarding unconditional, are then easily implemented on the intelligent network platform.

4.2 SERVICE IMPLEMENTATION ON INTELLIGENT NETWORKS

Services implemented on intelligent networks are not built in the traditional way. On the contrary, the intelligent network platform provides greater flexibility to service creation in general and also to the tailoring of services to suit the exact requirements of a particular customer.

This section examines each step of service implementation, like service creation, provisioning, and customization to the needs of the customer, and, finally, network introduction.

4.2.1 Service Creation on an Intelligent Network

Figure 4.3 illustrates a network with both node-based and terminal-based service implementations. Figure 4.4 incorporates IN-based implementations as well.

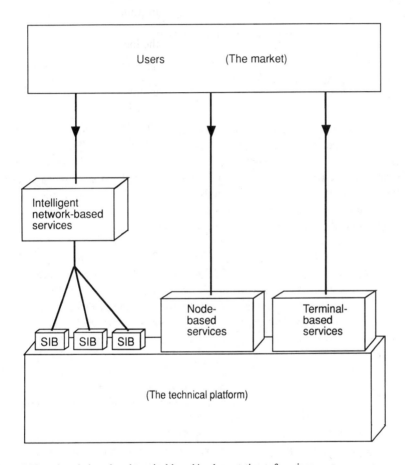

Figure 4.4 IN-based, node-based, and terminal-based implementations of services.

In the same way that node-based and terminal-based services are fixed and not so easily changed, as Figure 4.4 shows, IN-based services rely on service-independent building blocks (SIBs) that are fixed on the platform. Nonetheless, these services can be built from any combination of SIBs and are therefore more flexible. Again, the great advantage of IN-based implementation is the drastic reduction in time—from the two to five years required for conventional (node-based) programming and testing to only a few months. And, of course, this also implies that a service, when no longer wanted, can be easily removed or changed in a short time period.

As discussed before, SIBs can vary in size from system to system or even within the same system. Some systems contain numerous small SIBs and others contain only a few large SIBs. However, in service creation certain combinations of SIBs appear more often than others. It has therefore been convenient to form an intermediate level between services and SIBs, or the *feature level*.

Service creation on an intelligent network involves three basic components—SIBs, service features, and services—which are described below.

SIBs. SIBs are the smallest building blocks in service creation. They are fixed entities integrated with the stable intelligent network platform. The ITU Rec Q.1203/I.329 defines SIBs as follows:

> "A SIB is a standard reusable networkwide capability residing in the global functional plane used to create service features. SIBs are of a global nature and their detailed realization is not considered at this level but can be found in the distributed functional plane and the physical plane. The SIBs are reusable and can be chained together in various combinations to realize services and service features in the service plane. SIBs are defined to be *independent* of the specific service and technology for which or on which they will be realized" [1].

Service features (SFs). Some functions are commonly used for many services, for example, number translation services or mass calling services. These functions are called service features, or features. A service feature is built of one or more SIBs. Examples of features are time-dependent routing and origin-dependent routing. Features are often combined by using a set of SIBs, but they are not in themselves services, that is, they should not be used directly by users. They are combined with other features, SIBs, or both to create a service.

The ITU Rec Q.1211 defines features as follows: "A service feature is a specific aspect of a service that can also be used in conjunction with other services/service features as part of a commercial offering. It is either a core part of a service or an optional part offered as an enhancement to a service [2]."

Services. A service is something that can be seen and commercially offered to subscribers (users). A service is built by combining one or more SIBs and features or services or both. (I prefer to include services as well as SIBs. Services are not included in the theoretical model in ITU-T and ETSI, but reflect practical implementations in which new services are built by combining SIBs, SFs, and already-existing services.) In this way, simple services can be a subpart in an advanced service. (Please refer to the example, "Building a Simple Personal Communication System," on page 73.)

ITU-T Rec Q.1211 defines a service in this way: "A service is a standalone commercial offering, characterized by one or more core service features, and can be optionally enhanced by other service features" [2]. Consequently, we can extend the model in Figure 4.4 by adding the feature level. See Figure 4.5.

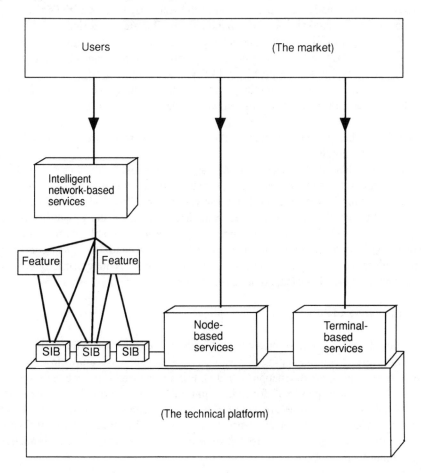

Figure 4.5 The *feature level* is added to service implementation.

Examples of services include freephone and televoting.

Service Creation Steps

The procedure for creating a new service can be easily divided into three main steps:

- Step 1 is the creation of a description of the service based on the user's perception of the way it should function. Normally, this step is best carried out by the marketing department of a network operator. Examples are descriptions of the general requirements for time- and origin-dependent routing possibilities and customer control functions.

- Step 2 is the creation of a description of the technical aspects of the service, including all error situations and how to handle them. This step is carried out by the technical department and is often done in *software description language* (SDL) according to ITU-T. SDL is often used because it is a high-level language (HLL), but the main reasons are that it is widespread, well known, and standardized.
- Step 3 is the implementation step, which entails loading the service logic into the SCP and updating the routing information in the SSPs and local nodes.

Once you become accustomed to them, steps 1, 2, and 3 can be carried out very quickly (in a few months).

In the very long run, service creation may be carried out by third parties on behalf of customers; that is, the service creation tools may be made available by network operators to customers, who may do the work themselves or hire consultants (third-party service creation). Whether this will become reality or not depends on the possibility of having, first, a *stable, well-defined intelligent network platform*, preferably with standard interfaces and application programming interfaces (API) and second, *extremely good tools* for testing services, for example, regarding service interaction. A new service, wrongly defined and not sufficiently tested, could harm other services and their users.

Service Creation Examples

This section describes some service creation examples based on ITU-T Q 1211 [2]. ITU-T rec. Q.1211 describes the mapping between services and service features in Capability Set 1 (CS1). A table presents the mapping, that is, it indicates which service feature is used in which service. (See Appendix A for a description.) Mapping is divided into *core* and *optional* features. Core features are features that are fundamental to the service, without which the service would not make sense as a commercial offering. Optional features, as the name implies, are not required by the service, and the service can be a commercial offering without them.

1. *Malicious Call Identification Service*

The malicious call identification (MCI) service is built by combining the following core features:
- *Call logging (LOG)*. Allows a record to be prepared each time a call is received by a specific telephone number.
- *Originating call screening (OCS)*. Allows the served user to bar calls (call screening) from certain areas based on the district code of the area from which the call originates.

No optional features are of interest here.

2. *Call Forwarding Service*

The call forwarding service is built on one core feature:
- *Call forwarding* (CF). Allows the user to unconditionally address incoming calls to another number. (It should be noted that a service and a service feature can have the same name.)

Call forwarding can incorporate two optional features:
- *LOG.* (See example 1.)
- *Customer profile management* (CPM). Allows the subscriber to manage the service profile in real time, that is, change answering places, control the announcements to be played, perform call distribution, change time-dependent routing, and so on.

Combining the optional feature CPM and the core feature CF creates the remote control of call forwarding unconditional service, the node-based implementation of which is described in Section 4.1.

3. *Automatic Alternative Billing Service*

The automatic alternative billing (AAB) service allows a user to call from any phone and have the call billed to a separate account instead of charging the calling or the called line. It is built by combining the following core features:
- *Authorization code* (AUTZ). Allows a user to get special calling privileges according to the authorization profile. Different sets of calling privileges can be assigned to different authorization codes, and a code can be shared by multiple users. The feature was first used in virtual private networks (VPN), to allow users within the VPN to override certain calling restrictions of a VPN station. Its use is expected to be broader in future.
- *Originating user prompter* (OUP). Allows the service to provide an announcement that asks the caller to enter a digit or series of digits via a DTMF phone or to enter a voice instruction via a more advanced service in future.

Two optional features may be added:
- *Abbreviated dialing* (ABD). This feature allows abbreviated dialing digit sequences to represent an actual dialing digit sequence; for example, a two-digit code may represent a complete directory number.
- *LOG.* (See example 1.)

Building a Simple Personal Communication System

To better understand how two services can be combined to provide more benefits than each service can separately, consider the combination of two of the services just described.

Combining automatic alternative billing and call forwarding with the remote control function results in a simple *personal communication system* (PCS).

A PCS is for individuals who want to be mobile (terminal-independent), both as a calling and a called party. This means that you, as a calling party, want to be able to call from any phone and be billed to your personal account instead of to the account related to the phone from which you are calling. AAB makes this possible. As a called party, you want to be reached wherever you are. CF with remote control offers this possibility. Whenever you move to a new location you can use the remote control feature to have your calls forwarded unconditionally to the terminal at that location.

4.2.2 Service Provisioning and Customization

When a service is built on the intelligent network platform from a combination of the right mixture of SIBs, for example, a freephone service or a credit call service, we say that a *basic service* has been created. However, different customers have different requirements for a freephone service. The needs of a pizza company with 100 restaurants differ from those of a bank with 20 offices. That is why we must adapt services to suit exactly the business needs of a customer. This adaptation is called *service provisioning and customization.*

As a result of provisioning, a certain number of features are available for use by a customer; in other words, a particular instance is created for the customer. Customization involves the customer tailoring the service to suit particular needs or, in other words, this is the customer's possibility to tailor the provisioned service to his particular needs. To do this, the network operator and the customer work together to identify the true needs of the customer. Let us return to the example of the pizza company offering pizza delivery service to its customers. Their goal is to have only one freephone number for the whole country (or the whole area covered by the company). Now we must find out:

- How should the total area that the pizza company covers be divided into subareas, so calls to a unique pizzeria number will be answered at the appropriate location? Which of the local pizzerias should cover which area? This depends on the capacity per local pizzeria, among other things.
- What are the hours of operation? What action should be taken when the pizzerias are not open (e.g., answering machines)? If there are different hours of operation for different pizzerias, should they cover for each other?
- What should be done when there is high traffic? Should the calls be divided automatically between two or more pizzerias in the same neighborhood? Or should there be automatic backup capabilities to a second pizzeria if there is no answer at the first (because it is not open or is experiencing high traffic)?
- Who should handle customer control, that is, who should be able to change the aforementioned parameters?
- What actions should be possible from the customer control interface? Should different users have different access to data? For example, should a local manager's ability to change data be more limited than the general manager's? Actions like enabling

and disabling customer control functions and determining the level of control local managers have, as well as managing passwords, should probably always be restricted to the general manager. (Customer control is described in more detail in Section 4.3.2 and is also discussed in Chapters 5 and 6.)

4.2.3 Network Introduction of Services

Once a service is created and customized, it is ready for network introduction. At first glance, this task might seem very simple, but remember that it is not only a matter of the service working technically. Additional systems must be updated, such as systems for administration, charging, and billing and customer databases, to mention the most important ones. At this stage, we must also distinguish between *implementing a new basic service* and *provisioning of an existing service for a new customer*.

When a new basic service is implemented, substantial effort must be devoted to integrating it into the network before customers can subscribe to it. This integration includes:

- Testing the service (software testing, functional testing).
- Loading the software onto the SCP or SCPs.
- Providing the service with a unique network access code so that there is an indication of when to call the SCP. The access code must be introduced onto the network, that is, the local and transit exchanges must be taught to transfer such a call to the SSPs, and the SSPs must be taught to recognize the prefix and call the SCP or SCPs dedicated to the service.
- Updating administrative, charging, and billing systems and the customer database.

If the basic service already exists and we are offering it for a new customer, there are other matters to consider. Service provisioning and customization involves:

- Giving the customer a unique number for this service in the number series for this service.
- Implementing the customized variation of the service in the SCPs, for example, with time- and origin-dependent routing.
- Enabling the customer control function, provisioning of passwords, and determining the level of actions that should be available for different users. (See Section 4.2.2.)
- Introducing the customer into the administrative systems, such as customer databases and charging and billing systems.
- Determining whether the service should be available for everyone, only for customers within the country, only from a certain region or regions, and so on.

Occasionally, the network operator may decide to implement a new service in a local region before introducing it nationwide. The operator might do this in response to the market situation or in order to explore a technical solution for the service.

From a *marketing point of view*, introducing a service in a limited geographical area before it is implemented in the entire network may be convenient. We can test the service in the local area before we decide on the next step. As we have not put too much effort into introducing the service, it is easier to withdraw it later on if it does not meet our expectations. Hopefully, however, we may decide to introduce the service throughout the entire network. To summarize, the market reasons for introducing a service in a limited area are:

- To see if customers use the service at all before deciding on wide-scale introduction.
- To limit risk. If unexpected misuse, abuse, or swindle takes place, the situation is easier to handle and the financial loss is less if only a limited area is affected. These types of incidents, if they happen at all, often occur very soon after a new service is launched.
- To test two similar services in two separate areas, to help network operators determine which one should be kept.
- Simply, to maintain different services in different areas.

A limited geographical introduction of a new service could also be *dependent on the technology*. For example, the current technical solution may have limitations in capacity that make it impossible to introduce it on a broad scale. Sometimes the "final" technical solution needed to introduce a service must be postponed because some vital components (SIBs) are still missing. Nevertheless, in cases like these we can sometimes get the service started by relying on a temporary solution. The temporary solution, however, may not have the same geographical coverage as the final one.

If we decide to introduce the service networkwide at a later stage and are planning to change the technical solution for it, our goal should be as follows: *From the user's point of view, it must be just as easy to use the service no matter what technical solution is used (backward compatibility). This means that, as far as possible, subscribers should not notice any technical changes to the network.* While this goal is easy to state, it is not always easy to follow. Impediments to the goal include: a lack of standards, which are required to produce compatible products; a lack of standardized products available from vendors to the network operator; a lack of harmonization, which is required to provide one service over more than one network (often, the user interface must be compromised).

4.3 THE FIRST IN-BASED SERVICES

The first services implemented on intelligent network platforms are generally the same on almost all networks. Many of them, however, began their lives as node-based services. This section is divided into three parts:

- Section 4.3.1 defines the *basic functionality* of some of the most common IN-based services in use today.
- Section 4.3.2 describes *value-added features* that can be incorporated in the basic services described in Section 4.3.1.

- Section 4.3.3 explains how *advanced IN-based services* can be created by combining features described in Section 4.3.2 with services described in Section 4.3.1.

4.3.1 Basic Characteristics of Some IN-Based Services

The following are short descriptions of the basic functions and characteristics of the most common intelligent network–based services in use today.

Freephone

Freephone (800 or green number) service is the "queen" of intelligent network–based services. The whole intelligent network concept is said to have started in the United States in the 1980s when Bellcore tried to find a good technical solution for the freephone service. This service is also known as the 800 service in the United States and in some countries in Europe, which refers to the number prefix used. Interesting to consider here is the fact that the 800 service, consequently, is older than the intelligent network concept.

The freephone service allows the calling party to call free of charge. The called party pays for the call. The service has become very popular because it allows a company with many customers to have all incoming calls charged to its own account. The main reason why companies use the service is, of course, to improve business volume. Typical freephone service subscribers include restaurants with food deliveries, banks, insurance companies, and airline companies.

The service becomes very powerful when combined with time- and origin-dependent routing and customer control. See the examples in Section 4.3.3.

Universal Number

This service works in exactly the same way as the freephone service, but charging is done in the normal manner (i.e., the calling party is charged). It is used when the called party does not want to pay for incoming calls but does want all the benefits of an intelligent network, such as time- and origin-dependent routing and customer control. Examples of the use of those benefits can be seen in Section 4.3.3.

Premium Rate with Shared Revenue

The premium rate (900 or kiosk) with shared revenue service allows the subscriber to call a certain number and get information, provided via the network, from an *external service provider,* in other words, from a provider who almost always is not the network operator. The call is charged (often at a higher rate than normal) to the calling party. The network operator collects the charge, keeps a part (related to the handling of the call), and transfers the rest to the service provider. The network operator handles the connection, the charging, and the transfer of money, but normally does not assume responsibility for the services delivered by the service provider. This is purely a business between the calling party and the service provider.

Service provisioning can either be direct, such as a lawyer or a doctor offering advice over the phone, or automatic, such as an answering machine that provides news, share prices, jokes, games, and so on. The premium rate service may be combined with time- and origin-dependent routing and customer control (see examples in Section 4.3.3).

Credit (Card) Call

This service, which is also called account calling, can be divided into two types, depending on whether a card is used or not.

The service *credit call (without a card)* allows you to call from any phone by providing your account number and a password. The account of the terminal you use will not be charged, only your own account. The advantage of this service is that you can use any phone capable of generating tone signaling (DTMF), allowing a wide use of the service.

The procedure for making a credit call is:

1. Enter the credit call access number, for instance, a freephone number, to reach the service.
2. Enter your account number.
3. Enter your (secret) password.
4. Enter the number of the called party.

Steps 2–4 consist of a dialogue between the user and the system whereby the user is guided through the sequence step by step. The major drawback of this solution is that you have to enter a lot of digits (about 35–45), which is a hard work and drastically increases the risk of making a mistake (entering the wrong digit).

With the *(credit) card service*, the terminal reads the card, automating steps 1 and 2, so that you only have to dial a password and the called number. A phone that is capable of reading cards is preferable and will probably be the dominant method in future, when card-reading telephones become more widespread, probably initially in public places. Today, however, there are not many phones that are capable of reading cards, hindering the expansion of this service in the short term. One way to solve this is to carry an automatic tone sender that will enable you to use any DTMF phone. You place the tone sender directly against the microphone in the handset to initiate steps 1 and 2. After that, you only need to enter the password, followed by the number of the party you are calling. Of course, this is only possible if the local exchange can detect DTMF pulses.

This aspect of the problem, namely, that DTMF phones are not widespread among subscribers and DTMF detection capabilities are often not available in the local exchanges, is true for many networks today. These networks still have older, dial pulse (DP) phones, which are not capable of sending dual tones to the network. To replace all these DP phones with DTMF phones would be a time-consuming and costly procedure; introducing the DTMF capability in a local exchange would be easier. An automatic tone sender could be the solution, provided DTMF capabilities exist in the local exchanges. If the automatic tone sender is equipped with a (manual) pushbutton facility that can translate numbers, the pound symbol (#), and the star (*) to DTMF signals, passwords and digits could be initiated and credit call and other services could be used from DP phones.

Alternate Billing

This is not a single service, but a group of services with the same purpose: to provide an alternative to the normal billing of the calling party by billing a *third party* instead. The third party can be another subscriber or an account. Alternative billing can involve:

- Automatically billing a third party for all calls;
- Automatically billing a third party for some predetermined B numbers only;
- Shifting billing to a third party in the middle of a call by using a code and a password.

Televoting

The basic idea of the televoting service is to use the telephone network for voting. Voting is a one-off event and must be announced in some way to make people aware of it. Consequently, television and radio are often used. However, a newspaper announcement can also initiate televoting. Games such as lotteries, in which subscribers can choose between two or more numbers by voting, are typical examples of televoting. Just as it is for the premium rate service, the cost may be somewhat higher than for a normal call (optionally, the network operator collects the money from the subscribers and gives a part of it to the organizer of the televoting). The organizer awards prizes to some of the subscribers and often allows subscribers to talk directly to the radio or television show host.

Virtual Private Network

Virtual private networks (VPN), which are actually more a concept than a service, allow a company that is spread out geographically to function as if it had its own private network. The private network is simulated through the use of certain functionality on the public network. Often, but not always, simulation is achieved through a combination of PABXs and functionality in the public network. A VPN provides unique number series, services, and so on, just as if it were a separate network. In fact, users should not be able to tell the difference between a VPN and a separate private network. They can, for instance, dial an extension number to make an internal call, just as if they were connected to a PABX. For external calls, they use a prefix to reach the public network, just as any call from a private to a public network would.

Although initially many VPNs were created outside the intelligent network platform, the advantages are so obvious that intelligent networks will probably become the dominant method of implementation in future. See also centrex and wide area centrex services, described below.

Centrex and Wide Area Centrex

In much the same way that a VPN simulates a private network for users, the centrex service simulates the functionality of a PABX (private exchange) in the public network. Again, subscribers should not be able to tell if they are connected to a PABX or a centrex.

An extension of the centrex service, the wide area centrex (WAC) service simulates the functionality of an entire private network consisting of several PABXs. Thus, the centrex service can be seen as a part of the WAC service in the same way that a PABX is a part of a private network.

Differences between the WAC and the VPN concepts (see above) are not so easy to find because the services, for users, are so similar. One implementation distinction is, however, that unlike the VPN concept, the WAC concept can not include PABXs.

Personal Communication Network and Universal Personal Telephony

The personal communication network (PCN) and universal personal telephony (UPT) are each concepts that contain a large number of services.

The basic function in a PCN is, of course, *personal mobility*. Customers (users) on a PCN receive a personal number on which they can be reached regardless of their location on the network. Moreover, they can use any terminal (any phone) in the network and have their calls charged to their personal accounts. A PCN includes a section for outgoing calls (A) and a section for incoming calls (B). The A section, as we saw above, allows you to charge your personal account instead of the phone you are using and is quite simple. The B section, which locates you wherever you are, is more complex.

While the PCN has yet to be fully implemented, some resolutions do exist that will gradually evolve towards the ultimate goal, UPT.

In traditional, fixed networks, a call is made to or from a fixed access point that is connected to a local exchange. When the first radio (cellular) mobile networks were introduced, the call was made instead to or from a terminal. This terminal was able to move around the network, without users having to consider that fact. In UPT, a further step is taken, namely, a call can be made to or from a person. The person can use any terminal or access point. By means of UPT, users can have this freedom regardless of the type of network. UPT is currently (1994) subject to standardization by the International Telecommunication Union-Telecommunication Standardization Sector (ITU-TS) and the European Telecommunications Standards Institute (ETSI).

In UPT, the network must be able to distinguish between the different individuals who use a certain terminal. This is essential, among other things, for charging purposes. This identification can be accomplished in several ways; but in groupe speciale mobile (GSM), for cellular networks, a removable personal card, called a subscriber identity module (SIM), is used in the terminal. Please, refer also to Section 6.1, items 8 and 9, as well as to Section 6.2, regarding terminal and personal mobility.

The type of mobility that PCN and UPT offer is personal mobility. Mobility is normally divided into two types: personal mobility and terminal mobility. The distinction between the two and their separate abilities to evolve in future, as well as the use of personal numbers, are discussed in more detail in Section 6.2.

4.3.2 Value-Added Features

Features, which are defined in Section 4.2.1, can be considered as large building blocks commonly used in more than one service. The following is not meant to be a complete list of features, because the main goal of intelligent networks is to provide flexibility—that is, to always allow invention of more sophisticated features and to create requirements for new SIBs. Here, we take a closer look at some common features simply to encourage an understanding of what features really are and to identify some of the interesting features in use today.

Call Distribution

Located in the *service control function* (SCF), this feature allows incoming traffic to be divided between two or more locations in a predetermined way. It is used when more than one identical resource is available and we want to divide the traffic load.

Time-Dependent Routing

This feature, also located in the SCF, allows different actions to be initiated depending on what time it is. The most common way to use this feature is with number translating services, such as freephone, whereby different C numbers (results from translation) can be given back to the service switching function (SSF) by the SCF depending upon:

- The time of day,
- The day of the week,
- The day of the month,
- The type of day (working day or free day) it is for the service number holder,
- Any combination of the above.

In the future, it will be possible to use time-dependent routing for purposes other than number translation services.

Origin-Dependent Routing

This feature, located in the SCF, allows different actions to be initiated depending on the origin of the calling party. Like time-dependent routing, this feature is primarily used for number translating services. Depending on the origin of the call, different C numbers can be given back by the SCF to the SSF.

Time- and Origin-Dependent Routing

This feature, which is a combination of time-dependent and origin-dependent routing, is a very powerful tool that is mainly used with number translation services. The feature allows networks to, for example, route calls from area 1 to answering place 1 and from area 2 to answering place 2 during normal working hours. At all other times, the calls from both areas might be routed to answering place 3. All this is handled by the SCF in the intelligent network.

Customer Control

This feature allows service subscribers and service providers to gain access to, read, and change *their own data*. Usually, only the network operator is allowed to do this. In its simplest form, customer control may be carried out from a normal DTMF phone. However, as the capabilities for making changes become more sophisticated, more advanced terminals will be required. On intelligent networks most customer control accesses will involve the service management point (SMP), which in turn updates the SCP, but simple changes can be made directly in the SCP.

Call forwarding or the handling of old, node-based services from a normal phone can be seen as the first step towards customer control, but the possibility of influencing service by those methods is very limited compared to what is possible on an intelligent network.

The most common uses of customer control in the early phases of intelligent networks will probably be in three main areas. First, the feature could be employed to control both time- and origin-dependent routing. For example, the feature could be used to change answering places, to change time-dependent routing for freephone or premium rate services, and so on. (See the example in Section 4.3.3.)

A major company with numerous credit call cards spread among its employees suggests a second area. Customer control access into the service management function (SMF) would allow the company to manage all of its facilities. The users (employees) themselves would also be able to make certain changes. (See the example in Section 4.3.3.)

A third area would be managing traffic flow, that is, routing management. Consider a major company spread out over a large geographical area, where the customer control function may be used to change the routing in the network between two subsidiaries located far apart. The telecommunications manager may use time-dependent routing and customer control to change the routing each time tariffs change, thereby ensuring that the most inexpensive routes in the global network are always utilized. Many networks have different charges at different times, for example, lower charges when traffic is low (evenings, nights, holidays). When these lower tariffs are in effect at different times (because of different time zones), a telecommunications manager can change the routing to save money for the company.

Security is a basic issue for customer control. Security is important to the owner of a service, especially where the risk of mistakes, abuse, and fraud is concerned. It is also important to the network operator, who is giving a third party access to the system in a way that was not possible before. Mistakes or deliberate misuse of the customer control function may affect the rest of the network, other services and service providers, and other subscribers.

Calling Line Transfer Capability

This general feature can be used in many different services. The most commonly used or discussed are the calling line identification presentation (CLIP) and the calling line identification restriction (CLIR) services. With CLIP, the calling line's identity, the telephone number (the A number) is forwarded to the called party (the B subscriber), where it can be used either for presentation on B's phone or for other purposes. CLIR is a service available to A subscribers who do not want their numbers to be forwarded to B subscribers.

Some B subscribers, however, do not want to receive any calls from individuals who have CLIR service, that is, individuals who do no wish to identify themselves. This disagreement has led to at least two schools of thought. One group feels that it is a violation of personal integrity to have numbers presented to the B subscriber; the other group thinks it is everyone's right to know the identity of a caller. There are also legal aspects in different countries that must be considered regarding restrictions for calling line transfer capabilities.

There is a positive spinoff effect, however. This feature makes it possible for the calling line's identity to be available in other parts of the network during a call setup. In future, it will be possible to use this very interesting feature in combination with numerous services.

Section 6.2.2 offers a closer look at future uses of calling identities.

4.3.3 Creating Advanced IN-Based Services

If we add the features discussed in Section 4.3.2 to the basic services described in Section 4.3.1, more advanced IN-based services may be created. Obviously, there are many combinations, and we cannot possibly cover all of them here. Therefore, I will focus on two examples that show how value-added features can be added to IN-based services, creating, in this instance, an advanced freephone service and an advanced credit (card) call service.

Advanced Freephone Service

Let us look at how the advanced freephone service, with sophisticated features, can be used in a real business by returning to the pizza company example described in Section

4.2.2. In that example, a major pizza company, located in a city, delivers pizzas within 30–60 minutes after a customer has called. The pizza company has 17 locations throughout the city.

First, consider the old, non-IN (non-advanced freephone) way, illustrated in Figure 4.6. A hungry customer must know where all the pizzerias are located and also must find the correct number for the one he wants to call. If he cannot get a delivery from one pizzeria, he has to look up the number and call another location. Each time, he must consider the location of the pizzeria, to ensure that it is not too far away. And he still may not be able to get a pizza delivered because the line is occupied or the pizzeria is too busy or closed, and so on. Another problem arises when an old advertisement and number remain in the telephone book for some time after a pizzeria has moved to a new location and changed its number.

Now, let us see what happens when the pizza business employs the advanced universal access freephone service available on our intelligent network. (See Figure 4.7.) The idea here is for the pizza company to have only *one* universal number, one number for all 17 locations throughout the city. (This is often, but not necessarily, a freephone number.) Instead of leaving it to the customer, the network is responsible for contacting the most suitable pizzeria. When a customer dials the unique number, he automatically reaches a nearby pizzeria. The network continues to try to reach a pizzeria until it succeeds. The order in which tries are made may be decided simply by a predetermined list or call distribution. In future, however, more sophisticated methods may be available.

Working together, the pizza company management and the network operator providing the universal number set up the service in the following manner:

- The 17 locations are divided into 5 regions, so that all customers within a region can always have a pizza delivered within 60 minutes from any pizzeria in that region.
- Customers use the same unique number no matter where they call from or what time they call. They no longer need to know the different pizzeria locations or when they are open, nor do they need to know the regions the pizzerias are divided into. They simply dial the "pizza number."
- When calls are received from a customer in one region, only the pizzerias in that region are potential deliverers of pizzas. By analyzing the customer's number, that is, the calling line number, the network determines the region from which the customer is calling. The network than uses a call distribution routine to allocate a certain percentage of the traffic to each pizzeria in the region.
- The general manager of the pizza company has access to the customer control function in the SMP. When a local manager calls to say two of her five employees are ill, the general manager can lift his phone and, using the customer control function, call the SMP, identify himself, give his password, and access his company data. He can now change the call distribution from, for example, 25-25-25-25% between the four pizzerias in the region to, let us say, 28-28-28-16%. After a couple of hours he will probably call the local manager to check that the workload is acceptable. When the staff is back to full strength, he can change the distribution back to 25-25-25-25%.

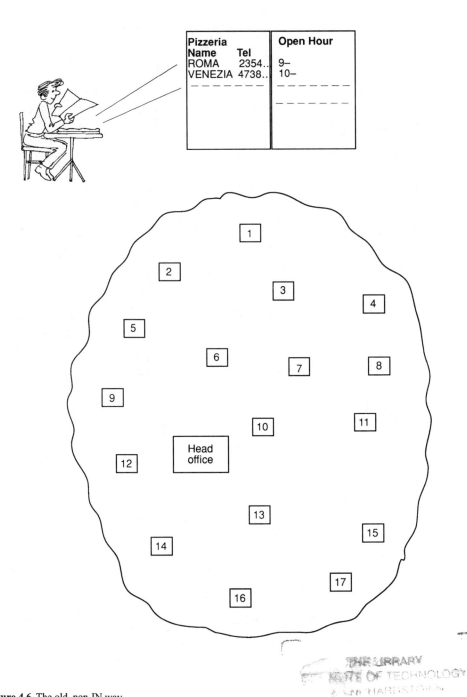

Figure 4.6 The old, non-IN way.

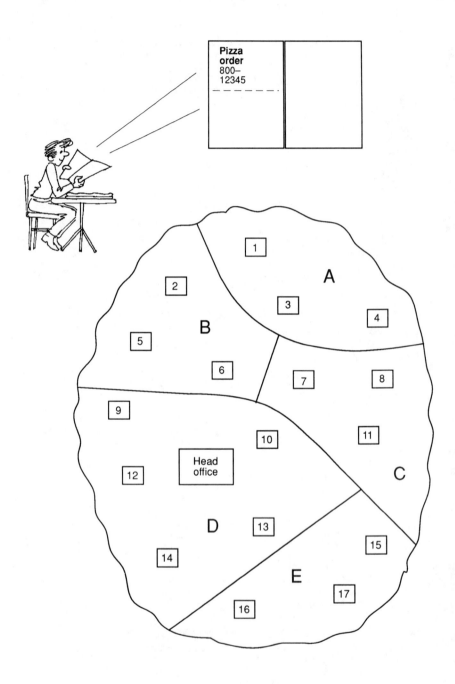

Figure 4.7 One unique universal access freephone number.

- If the pizzerias have different business hours, time- and origin-dependent routing can be used. For instance, 25-25-25-25% distribution may be valid until 9 pm. At that time, one of the four locations closes, but the other three remain open until 11 pm. At 9 pm the network can automatically stop routing calls to the pizzeria that closes and change the distribution to 33-33-33%. The manager, of course, could also use the customer control function to make the change.
- When all the pizzerias are closed, the network routes calls to an answering machine, which provides callers with information about hours of operation and so on.

Advanced Credit Card Call Service

As a second example of advanced IN-based services, let us consider a major company that has many traveling employees, each of whom uses a company credit card for telecommunication. Particular employees may have a card with no restrictions, a card that limits them to certain zones, or, in extreme cases, a card that only allows them to call one number (i.e., the head office).

As in the first example, the customer control capability is one of the most important facilities. The telecommunication manager is given full control over all the cards in the company. From the advanced terminal in her office, she can, among other things:

- Open new cards for new employees.
- Cancel a card when an employee leaves the company.
- Cancel a card that has been stolen or misused.
- Control the geographical range of a certain card, a group of cards, or all the cards.
- Control passwords for employee cards.
- Control cards that should only be opened at certain times, for example, only during office hours.
- Control the ability of a card to call only certain numbers, for example, forbid calls to premium rate numbers.

Employees also have customer control capabilities linked to their personal cards, enabling them to carry out a limited number of tasks, such as changing their personal password or temporarily canceling their card (e.g., to avoid misuse during holidays).

4.4 FAMILIES OF IN-BASED SERVICES

As we have already discussed, there are many similarities between the freephone, the universal access number, and the premium rate number services as far as implementation is concerned. They all have such features as:

- Number translating,
- Time- and origin-dependent routing,
- Customer control for changing answering places.

Shared features make these services practically identical *from a technical point of view*, though they are quite different from the user's point of view. In this section, we look at what we could call *families of services* (or service concepts) from the technical point of view. (Services can also be divided and structured according to similarities from the user's point of view, which is discussed in Chapter 6.)

Generally, when services are divided into families, a service can fit into more than one family. Nonetheless, from the technical point of view, it is possible to identify the following divisions of service families:

1. *Number translation services.* For these services, the SCF translates a called number to another number (hidden from the user), which directs the call to the intended party.

   Examples of services in this family include:
   - Freephone service
   - Universal number service
   - Premium rate service

2. *Mass calling services.* In this family, we find services that can create an overload on the network. They are usually characterized by the fact that they generate very modest traffic or no traffic at all most of the time. Suddenly, traffic soars to a very high peak. (See the televoting example in Section 3.5.) Traffic can be initiated by a newspaper or a radio or television program. Radio- and television-initiated televoting are potentially the most dangerous for a network because they are more *immediate* media forms than newspapers. When program audiences are asked to call a televoting number, traffic is initiated immediately. When newspaper readers are asked to do the same, the totality of the traffic could, in fact, be much greater, but it is more spread out over time and, therefore, the chance of overloading the network is not that great.

   Examples of services in this family include:
   - Televoting service
   - Premium rate service

3. *Alternate billing services.* Services within this family are billed in an alternative manner. In other words, some method of billing takes place other than billing the calling subscriber with a normal charge.

   Examples include:
   - Freephone service
   - Premium rate service
   - Account card calling service
   - Automatic alternative billing service
   - Credit card calling service
   - Split charging service

4. *Node- and network-simulating concepts*. In this family we find concepts like:
- Centrex
- Wide area centrex (WAC)
- Virtual private networks (VPN)

They are all described in Section 4.3.1 and have in common the fact that they wholly or partially simulate physical private functions on the public network. A centrex simulates a private automatic branch exchange (PABX) in the public network in such at way that users should not notice the difference. In the same way, a WAC simulates a network of PABXs. A VPN allows a mix of physical functions (real PABXs) and centrex functions.

4.5 COMPLEMENTARY SERVICES TO INTELLIGENT NETWORKS

As with most things in life, introducing intelligent network functionality on an already existing network is never simply black and white. Implementing intelligent network functionality on your network does not mean that all the services will become IN-based. In reality, you will have a situation in which many services are still node-based, some are IN-based, and some are terminal-based. Moreover, some will be new, some will be old, and some will be very old. No two networks will share exactly the same combination of service implementations. Consequently, if you want to provide a set of services to your customers as quickly as possible, the best solution will undoubtedly be to combine IN-based, terminal-based, and node-based services in a competitive set. Furthermore, even if we want to, we cannot modernize all the services at once because it takes time. Over the years, however, the manner of implementation will undoubtedly change, service by service, and as we discuss in Chapter 2, change will tend to favor IN-based and terminal-based implementation over node-based implementation.

REFERENCES

[1]   ITU-T Recommendation Q.1203/I.329, 1993.
[2]   ITU-T Recommendation Q.1211, 1993.

Chapter 5

Risks and Threats Facing Intelligent Networks

5.1 IDENTIFYING THE WEAK POINTS

The introduction of the intelligent network concept is, as we discuss in the preceding chapters, something of a revolution in that it makes it enables:

- Centralized control of services;
- Rapid implementation and customization of services;
- Remote control of services through direct access to the SMP;
- Expansion in the areas of business that can benefit from telecommunication services.

These are examples of how the intelligent network can improve areas of interest both for buyers (customers or subscribers) and sellers (network operators or service providers).

At the same time that these improvements will obviously yield many positive effects for the network and for users, they may also be *associated with risks.* If they are not carefully considered, these risks can pose serious threats to the future expansion of intelligent networks and, in the long run, the entire intelligent network concept. If the intelligent network is not introduced hand in hand with a careful examination of such aspects as security, availability, redundancy, tools for service interaction, and network planning we will undoubtedly encounter major problems.

The following sections describe a few of the problems that can arise.

All the Eggs in One Basket

Intelligent networks provide the capability of centralized control of services. This is a classic problem of putting all your eggs in one basket.

A service such as call forwarding or freephone, implemented in the conventional, node-based way is controlled *locally from many nodes,* each of which is responsible for the subscribers in its area. If the service is to be moved so that it can be controlled by the intelligent network, the first method of implementation that comes to mind is to let *a single*

91

SCP control the service for all subscribers on the network. This does not normally pose a capacity problem. What it does do, however, is concentrate all the control logic for the service in one location, that is, the SCP. If communication to the SCP breaks down or the SCP itself breaks down, the service will not be available to the whole network until the SCP is reached again or an alternative is established.

Centralization can also pose a problem for the businesses that use the service. Moving from a situation in which there was a dedicated number for each of a company's locations to a situation in which the company's entire business hinges on a single number places considerably greater requirements on the reliability of access to that number. This, in turn, places greater requirements on the technical and administrative systems that handle the service. (See the example involving a pizza company in Section 4.3.3.)

Too Rapid an Introduction or Customization of Services

Anything that is very easy to do and can be done very quickly always has risks associated with it. Consequently, we must consider the effects of all actions very carefully. This will undoubtedly be true when building and introducing services with the intelligent network concept. Too hasty an introduction of new services could prove to be costly. One way of avoiding this risk is to perform a careful test and validation of the test before implementing a service on the intelligent network platform.

Another phenomenon yielded by this relatively simple method of introducing services will be a much wider range of services available on all networks very quickly. On top of that, intelligent networks will enable us to customize services to suit special needs of customers. For example, a freephone service customized for a pizza delivery company with 100 locations will be different from that same service customized to suit a large insurance company. In addition, there are often different *service dialects* of one service or customized service. Service dialects are different implementations of one service that behave the same way as far as users are concerned. Service dialects are an important consideration, particularly when one is considering how different services on a network interact.

As we can see, the number of services, customized services, and service dialects on a network could multiply extremely rapidly. Service interaction and the inherent risk of undesired effects when two or more services are used together will become a major problem unless we develop good tools to handle this problem very quickly. In fact, the lack of good tools for service interaction risks may pose *the largest threat to the entire intelligent network concept.* Service interaction is discussed in more detail in Sections 5.2 and 6.4.3.

Another factor that must be considered as the number of services and service dialects increases is *careful management of access codes and number series.* Without management, we could quickly and easily find ourselves with no number available for a new service.

Complicated Customer Control Functions

Intelligent networks allow customers to control services remotely through direct access to the SMP. This control function is a lot more powerful than say, for example, the conventional control provided for initiating a node-based call forwarding service. But, while users can accomplish much more from the customer control interface, the damage they can do is also much greater.

Obviously, the first thing that comes to mind in this case is to limit what the customer can actually do. Managing passwords is a very important limiting device. Security must also be extremely high to prevent hackers from getting into the system and to minimize the risk attending mistakes. Organized fraud is another threat. A high-quality customer control system offers the best prevention.

Increased Reliance on Telecommunication

Intelligent networks allow organizations to expand the areas of business that can benefit from telecommunication services. And, as the possibility of using networks for such services as premium rate, freephone, credit calls, and alternative billing increases more and more businesses will abandon conventional methods of marketing and selling. Consequently, the financial loss suffered by a business if a network fails will become greater with each year. This places substantially greater responsibility on network operators and service providers to offer noninterruptable services as opposed to only basic telephony. In addition, as a growing number of businesses begin using telecommunication services, more and more money will circulate through the network. Undoubtedly, this will increase the risk of abuse and fraud.

Intelligent networks also bring about a marriage between the telecommunication network and such media as television and radio. Television programs, for example, use televoting or premium rate services to let audiences call directly during the television show. This often results in a mass-calling situation on the network, which can be very dangerous if it is not handled correctly. (Please see examples in Section 3.5 and Chapter 7.)

5.2 SOLUTIONS TO THE PROBLEMS

5.2.1 Redundant Paths and Backup Capabilities

Introduction of redundant paths and of another SCP (mated pairs of SCPs) capable of taking over control of a service for a customer without delay will be mandatory in networks. Normally, redundancy is required in case of a technical failure. In the case of a telecommunication network, however, redundancy will most probably be required as well to handle overload situations caused, for example, by mass calling services. (Mated pairs are covered in Sections 3.4.5 and 7.4.2. In addition, Chapter 7 takes a closer look at the concept of redundancy.)

It is important to note that the entire chain of systems comprising a network must be considered in the development of redundancy. In the first phase, establishing a secure and noninterruptable SSP-SCP communication is essential. The following configuration requirements must be put in place:

- Two SCPs physically separated with identical data at every moment that provide full redundancy should one fail. (Mated pairs of SCPs).
- At least two SSPs that can communicate with the SCPs.
- At least two separate physical paths between an SSP and the SCP.
- At least two separate physical paths between the local node and the SSP.

5.2.2 Local and Temporary Service Implementations

The ability to implement a service in only one region or to only selected customers offers a variety of benefits. First, if it is not clear how a service will be used once it becomes available, possible damage due to abuse or fraud will be minimized if only a limited number of subscribers have access. It is also easier to withdraw a service from the network if the service is implemented locally or restricted. Second, it might be that the service is only intended for local use. Third, some services may be designed for only temporary use, for example, credit card call and centrex services set up during a conference. Some services may only be seasonal, available during summer months or around major holidays, such as Christmas. Examples of seasonal services could include alternative billing and a freephone number that allows relatives to place calls without being charged.

To avoid running out of access codes and numbers, network operators may in future have to carefully coordinate their use. For example, two services that are not available during the same period or that are available in different areas could be given the same access code and number. It is essential, however, that network operators have very good systems for managing number plans. The intelligent network concept, which provides an overall view of the network, can be a valuable tool in this respect.

5.2.3 Fast Withdrawal or Temporary Closing of a Service

Intelligent networks allow services to be closed quickly at the SCP—for the whole network, certain areas, or selected subscribers. Closing of a service for a certain subscriber urges that we have the calling line number available in the SCP.

If we decide to withdraw a service from the network or temporarily close it, we must be able to do so *very quickly*. The ability to move very quickly minimizes the risk of losing money due to abuse and fraud for both the network operator and the service provider.

5.2.4 Built-In Control and Supervisory Functions

One way to minimize the risk of abuse and fraud is to introduce functions on the intelligent network that identify abnormal use of services. For instance, the SCP could initiate an alarm if

- An extremely high number of calls are made unexpectedly to a certain number,
- Extremely long calls are made unexpectedly,

- A credit (card) call owner makes two calls 100 miles from each other within 20 minutes,
- A card that was reported stolen suddenly appears.

The use of alarms must, of course, take into consideration normal user behavior for the service in question.

Supervisory tools such as alarms must be regarded as mandatory for network operators to avoid financial losses. However, legal aspects in different countries may create obstacles to this kind of supervision.

5.2.5 Service Interworking Management Tools

Tools that can manage service interworking, that is, the way services work together when combined, are mandatory if networks are to expand the number of services, customized services, and service dialects they can offer. Two or more services may

- Interact in a *negative* way,
- Interact in a *positive* way,
- Be impossible to combine simply because they are *alternatives* to one another,
- Be of *no use* at all when combined because of their functionality.

Some combinations have already been foreseen by the network operators, but far from all. Network operators know to forbid the use of certain services when an incompatible service has been invoked. For example, they do not allow users to initiate the premium rate service during a freephone call.

Negative service interaction, that is, when two or more services used together yield a negative instead of a positive result, is the most urgent matter to address. As this is an extremely complex and difficult problem and the combination of services and networks is enormous, we should not primarily attempt to solve the problem. Instead, the most important and urgent matter is to thoroughly understand it so we can direct our efforts in the right places. Service interworking, including service interaction, is discussed in more detail in Section 6.4.

Chapter 6

New Service Demands in the Future

Well, if I called the wrong number, why did you answer the phone?
James Thurber, 1894–1961 [1]

From this chapter onward, we take a step into the future. Of course, using the word "future" in a book is dangerous in many ways. I have no way of knowing *when* you will actually read this chapter. Moreover, there are no similar reference points to start from when comparing different networks, and there will never be. Networks will always differ, largely because of their sizes in combination with different generations of technique. Still, this is part of the charm of working with telecommunications. Nonetheless, we can identify some general trends in networks of the *future* that will result from discussions and studies of the existing situation in the 80s and early 90s and that have not yet become 100% real.

I am thinking of

- A wider scale of broadband communication,
- Mobility for narrow and broadband communications,
- An expansion of facsimile and voice mail,
- A wider range of services,
- Customization of services,
- A shorter time-to-market interval for new services,
- Increasing interworking between public and private networks,
- A wider use of paging systems.

However, before we can seriously introduce new services on a broader scale or at least in parallel, there is another more urgent matter to consider. We must first improve the number of successful communications, that is, increase the possibility of a call reaching the called party. Studies have shown that far less than 50% of all call attempts are successful.

(Unsuccessful call attempts do include instances of the wrong person answering the phone.)

The number of unsuccessful call attempts has probably risen during the 70s and 80s largely due to changes in our lifestyles:

- There is less probability today that someone is at home during the day, because often both man and woman work.
- We are more mobile at work and in leisure hours. More employees move around between different locations than in the 50s and the 60s. We spend more time than we did 20 years ago on leisure activities outside our home, like sports, education, weekend cottages, and so on.

How then can we increase the percentage of call successes? The solution can be divided in two main parts. First, we must increase the possibility to be mobile for users on the network, that is, either make it easier for an individual's communications to follow them when they change locations or at least use a method to locate them quickly. These steps are best accomplished by using services and network concepts that provide mobility. Examples of such services include call forwarding, time- and origin-dependent routing, customer control services, and paging systems. Cellular networks provide examples of network concepts. Combinations of these are, of course, possible as well.

Second, we can increase the number of successful calls and make it easier for calls that would otherwise fail to be completed by using the methods described below.

Some of the most common reasons why calls fail are:

1. The called party is busy,
2. The called party doesn't answer,
3. There is congestion on the network,
4. The calling party makes mistakes in dialing,
5. The wrong party answers the phone (colleague, husband, or wife).

These reasons for failed calls can be addressed by these correspondingly numbered methods:

1. When the called party is busy, use a value-added service, such as call waiting or call completion to busy subscriber, or send a facsimile.
2. When the called party doesn't answer, use a value-added service, such as call completion when no reply, or send a facsimile or voice mail message.
3. Not every call on a congested network is of the same importance. For example, an emergency call to an ambulance is far more important than a call to order a pizza. In the latter case, waiting a few more minutes is often of no importance. A priority function in the network could force less important traffic to wait during periods of high traffic, which would guarantee a higher percentage of successful

calls for high-priority traffic. To be powerful, though, priority must apply throughout the entire system. (This is further discussed in Section 7.2.)

4. Mistakes in dialing are often caused by a bad user interface and a lack of guidance. A more user friendly interface (Why not place a help button on the terminals in the future like we have on our personal computers?) would avoid many of today's mistakes and problems. (Please, refer to Section 6.1, item 10.)

5. The problem of the wrong party answering the phone can be solved by, for example, a distinctive (personal) ring or through the use of a mobility measure, where called parties plug a personal card into the phone they are near so their calls can find them. Introduction of personal numbers in future will also be of value in this instance.

It is also necessary to make a stronger effort to spread information about the benefits of the use of value-added services in general. In other words, we need to encourage greater user willingness. User willingness is discussed in Chapters 1 and 8.

In this chapter we discuss how demand for new services and groups of services will rise, taking into account new trends in society and networks. Grouping of services can reflect either the point of view of users or the network. Here, we consider the grouping of services from the users' viewpoint. Our examples include alternative billing services, services with time- and origin-dependent routing, and services with special demands for high security. We cover the future trend for new services in Section 6.1. Section 6.2 focuses on full mobility in the networks and how this impacts services in future. Service management and administration, including charge determination and billing, service prototyping, customer control, security, privacy, and reliability, are dealt with in Section 6.3. Finally, in Section 6.4, we discuss one of the most important matters in the future expansion of services, the interworking of services. Interworking is becoming increasingly important as the number of services, service dialects, and customized services grows.

In Chapter 7 we consider the impact these new service demands will have on future intelligent network platforms and telecommunication networks in general.

6.1 TRENDS, TENDENCIES, AND DEMANDS IN THE 90s

Trends and tendencies in society in the 90s will more or less be the result of the telecommunications advancements that intelligent networks offer. Examples include:

- A tendency to work more at (and from) home. A demand for the freedom to choose and change places of work.
- Demands for recreation, available at home and during leisure times. These include games and competitions, often in combination with television or radio programs.
- Demands for more immediate access to direct information, both in professional and private lives.
- Demands for secure and fast communication, both in professional and private lives.

In telecommunications, these demands will necessitate introduction of:

- *Basic value-added services*, to provide mobility, that is, the freedom to alter physical location.
- *Information services* (including interactive services) on telecommunications networks, which will begin to compete seriously with newspapers, radio, and television.
- *Customized value-added services,* tailored to suit special business needs. Value-added IN-based services are already mandatory for many companies. This trend will become increasingly important for all types of business over the years. Examples include virtual private networks (VPNs) and televoting.

The *consequences* for telecommunications networks will be that:

- More and more money circulates via public networks.
- More IN-based services, including customized services and service dialects, will be introduced.
- There will be greater focus on the growing number of services and the fact that customers will have different sets of services in the future, resulting in greater attention to the administration of services and service interaction.
- Demand will rise for a more flexible charging and billing system.
- Demand will rise for a faster time-to-market period for new services.
- Demand will rise for support services that provide mobility and remote control, for example, customer control.
- More efforts must be devoted to security, privacy, and priority aspects. This also includes a differentiation capability. Different levels of security, privacy, and priority will be available for different users (subscribers) depending on demand and, certainly, cost. High priority can ensure a fast and reliable (no congestion) connection through the network. High security could also include low bit error rate.
- Greater demands will be made for flexible bandwidth (bandwidth on demand).
- Demand will increase for databases (government and commercial), both in business and at home. At home, users will want to rely on knowledge databases instead of reading books when cooking, translating, searching for words in dictionaries, and so on. Knowledge databases may also be used for studying, also interactive. These types of services can employ a flexible approach to charging by providing access through a premium rate number.

Where services are concerned, this will result in such demands as:

1. A much wider range of services;
2. One business area = one service package;
3. Demand for both global and local services;
4. Increasing mobility;
5. Increasing use of non-real-time communication;

6. An increasing demand for wireless terminals;
7. Increasing security and reliability;
8. Increasing demand for identification services;
9. Demand for more use of cards;
10. Increasing user friendliness and simplicity of services;
11. Use of services with both narrow and broadband accesses for calls and services;
12. Increasing use of multimedia services.

Later in this section, we examine the demands in the service area one by one:

1. *A much wider range of services.*
   Where value-added service repertoires in fixed networks today are concerned, we will see an increasing number of new services in the 90s, as well as adaptation of existing ones to support mobility. We should also expect a formidable explosion in the use of facsimile and voice and computer mail via the network.

   We will also see a much wider range of service dialects and customized services in the networks in future. Very soon the number could reach 500–1,000 or more in a network. This will force us to place greater focus on matters like management of services, service interaction, and so on. An ordinary customer will not, however, use more than perhaps a maximum of 50 services, for practical reasons. A greater number of services than about 50 is not only unnecessary, it is impossible to handle. Different customers, and more importantly, different customer categories will have different demands, which will probably result in different basic service packages that are tailormade for different categories, such as:

   • The large company with many branches;
   • The self-employed, highly mobile person with one office;
   • The household.

   Moreover, special requests for additional services can be accommodated.

   The growing number of services will force us to focus on the way services are implemented and especially on the location of the service logic. (This is covered in Chapter 2 and in Section 7.5.) Do you remember the discussions about implementations in Chapter 2, Section 2.2.3? More and more attention will have to be paid to providing "the right service at the right place at the right time."

2. *One business area = one service package, that is, customized (tailormade) service packages for different areas of business.*
   In future, when we have access to many different node-, terminal-, and IN-based services, we will find out that only some of those services will be of interest to a specific customer. For example, large companies could be exactly the same size, but demand different service packages. Insurance companies, travel agencies, food delivery companies, and department stores could all have very different requirements.

3. *Demand for both global and local services.*

One trend in service provisioning is to make many existing services more generally available and to continuously expand the range of a service to include:

- Increasingly larger geographical areas;
- More than one operator's network;
- Different access networks, for example PSTN, ISDN, mobile.

On the other hand, there is also a trend to provide customers with services used in a very limited geographical area (for example, only where the customer practices his or her profession). These services are often customized to suit the needs of that single customer.

These trends contradict each other. Or do they? Both trends are real, but they each involve different types of services and are for different types of customers. Compare, for example, a pizzeria and an insurance company. The pizzeria bakes and delivers pizzas (works with a short timeframe) and must take a *local approach* to its business. The insurance company, on the other hand, does not have this geographical limitation or work with short timeframes.

Completion of call to busy subscriber (CCBS) is an example of a service that, without the limitations of networks and access forms, would be very useful globally. Televoting offered by a local radio station or a VPN service offered temporarily during a conference, on the other hand, are services that would be useful at the local level. Of course, a customized service, such as VPN, could also be used globally, depending on user requirements.

4. *Increasing mobility.*

We will experience a growing trend towards full mobility in fixed networks that will be much the same as the trend in the 80s toward radio-based mobile networks. We want to have the freedom to be independent of a physical (fixed) connection at a fixed location, both as a calling and as a called party. We also want to be able to work via remote access to the network, at home and so on.

We are most likely heading towards, in the future, full mobility in all networks. I believe that there is common agreement on that fact. But what does that mean? We must remember that full mobility is not something we can just decide to implement on our networks on a fixed date and then begin planning to accomplish. Instead, we achieve full mobility gradually by introducing new services and new technology. Many networks already have partial mobility, for example, with the call forwarding service for control of incoming calls. The problem is that we must return to our own telephone each time we change our location. From a technical viewpoint, for outgoing calls we can use any phone, but, of course, that does not take charging into account. Consequently, the ability to be mobile as a calling party depends heavily on charging flexibility.

So, instead we must start improving mobility by upgrading existing services and making them more user friendly and available at the places where we need them. A valuable existing service in many networks is the paging service, which lets us know when somebody is trying to reach us. All we have to do then is call either the number shown on the *pager* or a predetermined number (home or a secretary).

But what has been discussed so far are only the first steps towards full mobility. We must also plan for a more general mobility concept, by which we should be able to distinguish between different types of mobility, such as personal or terminal mobility. As mobility continues to grow, it will undoubtedly fuel the greatest revolution in our service repertoires by introducing new services and changing many existing services. It will also make many of today's services, such as call forwarding unconditional (CFU), obsolete. The CFU service we have today, from one access point to another on the network, will instead be split into different variants. First, for example, it will be an integrated basic function within full mobility between access points, and, second, it might be a "CFU" service that allows us to forward our calls to another person or function. (Mobility and the effects it will have on future services is further discussed in Section 6.2.)

5. *Increasing use of nonreal-time communication.*
The use of *facsimiles* (faxes) has really started to grow in the few last years. Some experts forecast that in 10–15 years, the use of facsimile will be as common as the use of a phone today. Within companies (often larger ones), the sending of *electronic mail* between personal computers (PCs) is one of the most common ways to communicate. A third example of non-real-time communication that is often used between people within the same company and that is expanding every year is voice mail boxes. These services are no doubt expanding because they are more attractive compared with the many frustrations one can experience trying to make a normal call. Also, people can send facsimiles, electronic mail messages, and voice mail at times when they can not normally make phone calls, for example, late in the evening, on weekends, and on holidays.

Of course, faxes and electronic mail offer other benefits, compared with speech calls, such as enabling called parties to receive a message at their fax machine or printer. Other benefits include the capability to send a multicast message, that is, a message with more than one receiver; automatic retries; and confirmation messages. One of the greatest benefits of these services and a driving force for increasing their use is that they do not depend on real-time communication, that is, they do not require the called party to be present during communication. This dramatically increases the number of successful calls.

My personal feeling is that these three services, facsimile, electronic mail and voice mail, have so much in common that they will converge and follow the same line of development in future.

6. *An increasing demand for wireless terminals.*
Phones with wall connections will eventually disappear and be replaced by portable phones. This will happen first at the office and later in the home. Each

member of the family could have a portable phone and be connected to the same radio-based station situated in the house. Portable phones can be used inside the house and outside in the garden. Due to their limited range, however, these portable phones will probably still be connected to the fixed network.

7. *Increasing security and reliability.*
The use of telecommunications for business and financial transactions (for example, home banking) will continue to increase dramatically during the 90s. This, together with the increasing use of telecommunications in combination with public media (like television) that require *real-time operation without interruption*, will make the networks more important—from an economical point of view—than ever before. Consequently, the demands for security and reliability will be greater each year.

Security and reliability are discussed in more detail in Section 6.3.4.

8. *Increasing demand for identification services.*
A good identification method (for example, cards, chip cards) will most certainly be required in future when we become more mobile and more and more financial transactions are carried out via telecommunication networks. These cards can have many uses, such as when identification is combined with a customer control function, during a credit card call, or when the calling party's identity must be given before or during a call. Of course, for security reasons, the identification card must be used in combination with a secret password.

In group mobile speciale (GSM), used in cellular networks, a removable chip card is inserted into the terminal for identification purposes. This card is called the *subscriber identity module* (SIM).

Identification and identification aspects are also covered in Sections 6.2.2, 6.3.2, 6.3.3, and 6.3.4.

9. *Demand for more use of cards.*
The use of all types of cards for telecommunications will increase during the 90s. Examples include:
- *Prepaid cards:* cards you buy in public places and use on public phones until the value is used up. When a card is used up, you just throw it away and buy a new one.
- *Credit cards:* (issued by network operators and public ones like VISA, Eurocard, and so on).
- *Identification cards:* used mainly for security reasons. The cards will probably be used first with a credit card service. (See item 8).
- *Chip cards:* these cards are removable and can therefore be used in any terminal. They are often used for both identification and charging purposes in concepts for personal mobility, like UPT. (See item 8.)

10. *Increasing user friendliness and simplicity of using services.*

As Chapter 1 mentioned, in a discussion about user friendliness and willingness, the first step towards a broader use of services is a better user interface. The most common terminals, which make use of ten digits, the star (*), and the pound sign (#), are very primitive. The next step is a terminal that carries the name of a given service on the corresponding button—one service, one button. Naturally, the number of services listed on the terminals will be very limited.

Future terminals will likely be more functional and house the intelligence for many services. Probably, they will be programmable for multipurpose use—users will be able to simply program their terminals according to their own needs. For example, users can assign the services they use most often to their terminal keys. The terminals will also probably have letters instead of numbers, for easier communication. Users will dial a name or a number or perhaps either. A dialed name will be translated into a number in the same way that we translate an abbreviated number today. The translation will be done locally by the terminal, the local exchange, or the SCP. Personal numbers and numbers that users dial frequently—spouse, relative, bank, dentist—can be programmed directly into their terminals. Official numbers, such as municipal or central government authorities, can be located on terminals, the local exchange, or the SCP. (The right number at the right place.)

Another highly interesting service in future will be a help button and help menus on your terminal that will be much like the help button on personal computers today. If you do not understand how to use a particular service, you simply push the service button in combination with the help button to receive assistance from the network, either by voice or by text (already possible today in a limited way). You will also be able to use the help button during a call or in the midst of using a service, whenever you encounter a problem.

The wide range of services that will be offered to subscribers will necessitate simple manuals. Even our present "relatively stable and limited" service supply causes problems for people (fear of technology). And to further increase user friendliness, we must also take care when making technical changes to the network. This is when *backward compatibility,* from the user's point of view, must be taken into account. As far as is possible, a subscriber should not have to learn a new way to use a service simply because the network operator changes technical conditions on the network.

11. *Use of services with both narrow and broadband accesses for calls and services.*

Narrowband services will certainly continue to dominate during the rest of the 90s. Many new services for speech will see daylight as network capabilities increase.

We will probably see a distinction in the evolution of services in different areas and with different evolution lines in future. First, we will see a distinction between: the business area and the private area. Later on, these areas will further divide into:

- Services that are suitable for different sized companies—from large, multilocation companies to small one-person companies—and specialized services for, for example, pizzerias, insurance companies, and banks.
- Services for social purposes (communicating with friends, official authorities, and so on), information (weather forecasts, news, and so on), and recreation (games, and so on).

Broadband services will undergo a similar division. The intelligent network, created first for narrowband services, may also be used here, that is, you may in future use freephone services between users with broadband access. Another future use will be premium rate service with broadband access, where you can rent movies and the like from your own home by using this IN-based service. You are charged per view or per hour directly on your telephone bill.

In future, when a call is set up between users with different access forms (for example, between a narrowband "ordinary" subscriber and a broadband subscriber), the SCP on the intelligent network can help find the common set of services you can use during the call.

12. *Increasing use of multimedia services.*
   In future, new terminals and new services will be available which may not have the same bandwidth requirements. Some may be for speech only, like today's phones. Some may be for text, some for pictures, and so on, and some may be a combination. During a connection through the network, it will become necessary to use different bandwidths, which means that we will be able to get bandwidth on demand, that is, demand the appropriate bandwidth for each call.

6.2 TOWARDS FULL MOBILITY IN THE FIXED NETWORKS

6.2.1 A New User Structure

Increased mobility, both for the calling and the called party, will be one of the major reshapers of services. When mobility is introduced in the fixed networks, the calling and the called party will no longer involve only the access points, as before. Instead we must structure them into more than one party, considering all four functions described below. Many new services will be introduced to meet the needs spawned by mobility. Some of the services existing before mobility will disappear because they are no longer needed, at least not after full mobility becomes available. (Please, see Section 6.1, item 4). Others will be changed or split into additional services because of the new structure of the calling and called parties.

We can identify four functions of interest for the calling and the called party regarding a call (Figure 6.1):

- The person who calls or is called (the user);
- The terminal used;
- The access point to by which the terminal is connected to the network (*network access point* (NAP));
- The subscription owner, that is, the person who pays.

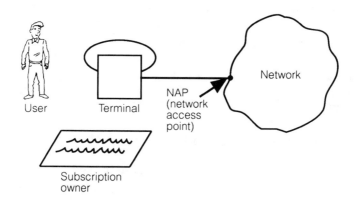

Figure 6.1 The four functions of interest in the introduction of mobility.

Let us investigate the impact of increasing mobility, step by step, and see how the mapping of these functions gradually changes.

Mapping before Mobility Is Available

A user (a person) in a fixed network, prior to the introduction of mobility, is bound to a certain terminal, for example the one at home. (Of course, there could be more than one at home but this is of no interest here.) This terminal is bound to a certain NAP, and this NAP, in turn, is bound to a subscription to which all outgoing calls are charged. See Figure 6.2, calling party.

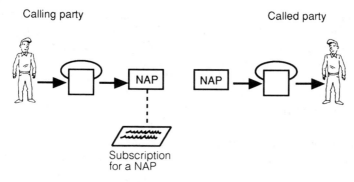

Figure 6.2 Mapping prior to mobility.

For incoming calls to a person, the opposite is true. We make a call to an access point, which is bound to a terminal that belongs to that person. See Figure 6.2, called party. Each change to a new terminal connected to another access point requires the opening of a new subscription. Consequently, what defines the subscription is the NAP. Please observe that if the freephone service is used the subscription is bound to the NAP of the called party instead. Nothing else changes. In other words, the subscription is bound to the NAP no matter which party is charged.

Mapping If Personal Mobility Is Introduced

When personal mobility is introduced, we are free to move around in the network and can make outgoing calls, on our own account, from any terminal. Compare the calling party in Figure 6.3 with the calling party in Figure 6.2. The differences are that

- A subscription is bound to a person and not to the NAP,
- A user is no longer bound to a certain terminal,
- A terminal in the fixed network is still bound to a certain NAP.

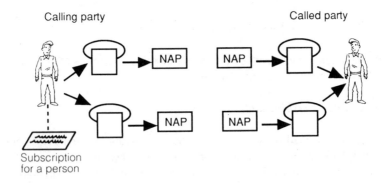

Figure 6.3 Mapping when personal mobility is introduced.

For calls received by a mobile person, the opposite is true. Calls go to different NAPs depending on where the person is. Each NAP is bound to a terminal. The person may receive the calls on different terminals. See Figure 6.3, called party.

One example of personal mobility that is outgoing is the use of the credit (card) call service.

Mapping If Terminal Mobility Is Introduced

If, instead, terminal mobility is introduced, the terminals are mobile and can be connected to any network access points. Users employ the same terminal for outgoing calls no matter where they are on the network. Compare the calling party in Figure 6.4 with the calling party in Figure 6.2 and in Figure 6.3.

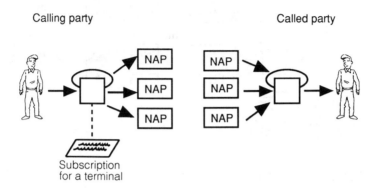

Figure 6.4 Mapping when terminal mobility is introduced.

The following difference, when compared with the situation before mobility, should be noted:

- A subscription is bound to a terminal and not to a NAP,
- A terminal is no longer bound to a certain NAP,
- A person is still bound to a terminal.

For incoming calls to a person using a mobile terminal, the opposite is true. Calls go to different NAPs, depending on the location of the terminal. A terminal may be connected to different NAPs, but a person is bound to his or her terminal. See Figure 6.4, called party.

The use of cellular networks for mobile phones, introduced in many countries in the 80s, is an example of terminal mobility. Here, however, we consider solutions that are also applicable in fixed networks.

Mapping If Both Personal and Terminal Mobility Are Introduced

If both personal and terminal mobility are introduced, we will have a situation similar to the one in Figure 6.5—a type of ultimate mobility with complete freedom. A user can choose any terminal for outgoing calls, and the terminal can be connected to any access point on the network. The subscription could then be bound either to the person or to the terminal. (Please compare with the calling party in Figure 6.2, Figure 6.3, and Figure 6.4.)

For incoming calls to this person, the opposite is true. Calls may go to different NAPs and the person may receive the call on different terminals. In other words, the terminal may be connected to different NAPs, and the person may use different terminals. See Figure 6.5, called party.

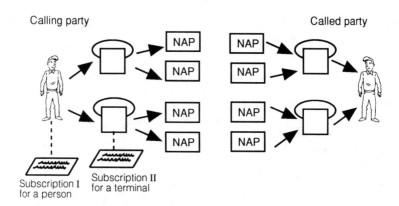

Figure 6.5 Mapping when both personal and terminal mobility are introduced.

Full Flexibility for Charging

The next step in the evolution is to allow total freedom to change the way calls are charged (see Figure 6.6).

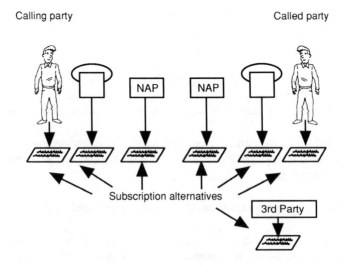

Figure 6.6 Total freedom to charge when full mobility is introduced.

Calls can be charged in different ways. They can be charged to:

- *The person who calls.* If a private person owns the subscription, he or she might be a private person who is mobile and could use any terminal in the network. Other persons could also be users of the same subscription, for example, in a family.

- *The terminal used.* If the terminal owns the subscription, it might be a company that has mobile terminals for its employees. Employees might take the terminals with them while working, but the bill goes to the company.
- *The NAP used.* Once full mobility is introduced, there will still in some cases be a need to charge according to physical location, that is, according to where the terminal is connected to the network. In this case, the conventional method of charging the access point is used. An example might be a company's connection to a network to which several terminals can be connected.
- *A third party.* This includes all other ways to charge a call. Examples of a third party include an account, another person, or another subscription.

In the first three cases, it should also be possible to charge the called side, that is, the person, terminal, or NAP called. Furthermore, it should be possible to split the charge between the parties involved.

2.2 The Impact of Full Mobility on Services

Moving towards full mobility means we must continuously change the services we have implemented, for each step we take. Mobility will probably be one of the greatest revolutions for value-added services since their creation, as it deals with almost every service that existed prior to mobility. In this section, we will examine some of the most important services available in networks today, services planned for the future, and services soon to be implemented to see how they fit into a network with partial or full mobility.

Calling and Called Identification Services

Of course the calling party identification service comes to mind first, as it is similar to the calling line identification (CLI) service that exists today. See Section 4.3.2. This service, which identifies the calling party prior to call setup, offers, among other things, more security to the called party.

Another identification service that will probably be of interest in the future is the called party identification service. But, wait a minute, don't we know who we are calling? Let us consider an example. If I call someone to discuss business, I want to be 100% sure that the person on the other end is the person I wish to speak with. If I call a bank, I want to be sure that I reach the right department, regardless of who answers. Well, you might ask, Do you know what number you dialed? Obviously, I know what number I dialed, but I do not know where I have been connected. The bank could have two, three, or more numbers for the office (a normal number, a freephone number, and so on), and these numbers could be used together with: time- and origin-dependent routing; customer control for routing the number to different places; call forwarding (unconditional, on busy, or on no answer); and so on. These variables make me highly uncertain of which physical location I have been connected to. Introducing a called identification service could minimize the risk of abuse and fraud, among other things.

Identification Cards

One way to identify yourself as a calling or a called party is to use a personal card in combination with a password, the same way you do with an automatic teller machine at your bank. The credit card service was probably the first type of service that used identification by card in telecommunications.

The Calling Party

Section 4.3.2 describes the calling line identification (CLI) services CLIP, for presentation of calling line on the called subscriber's terminal, and CLIR, for restriction to it, prior to the introduction of mobility. The ability to identify the calling party will be one of the most important features also when full mobility is introduced. However, we cannot talk simply about "calling line identification" anymore, as this is only one of the cases. Instead, from now on we must talk about the *calling identity* (CI) to cover all cases.

CI could be used for presentation at the called party, as in the CLIP service, but it could also have other, more sophisticated uses. In future, it could be available on different parts of the network during set up of a call or during a call. A good example is when both you and the person you wish to call move around on the network. During the call setup and the call, it is extremely important that the your two terminals can handle your call and the services you intend to use during the call. It is also important to check the two NAP against one another, because you could be connected to exchanges with different capabilities. A third example could be that you have two different personal service profiles. Perhaps one person has a low-budget set of minimal services.

So, the CLI service Section 4.3.2 describes is only valid when we have a one-to-one connection between the subscriber and the terminal (the phone) and a one-to-one connection between the terminal and the NAP. But, as we see in Section 6.2.1, when we enter the world of full mobility, this will no longer be the case. A person will be able to use any terminal for calls. Moreover, it is possible to move a terminal and use it in different places.

Which number, then, should be transferred for use on the network or for presentation at the called side—the caller's personal number or the number of the terminal? Or should it be an entirely different number? The answer is that in future the CI will be divided into at least five different identities, depending upon the actual use. So, from now on, we consider as relevant:

1. The calling person,
2. The personal function,
3. The terminal identity,
4. The terminal function,
5. The physical location, that is, the line or NAP (network access point).

The customer control function will enable added-value service capabilities to move between these five identities between calls or during a call.

Let us use a few examples to examine this new structure of the calling identity, or perhaps, we should say, the *calling party*:

1. *The calling person.*
   When I call my spouse, children, parents, and so on, I could, in future, receive special privileges (such as priority) compared with other callers. When I call my insurance company, they could use my identity, which is transferred with the call, to immediately find my personal file on their computer and give me a quick reply to my questions. This personal identity will be independent of the terminal and the NAP used.

2. *The personal function.*
   For example, the general manager of a company might call a bank or local or government authorities and receive special rights and access capabilities into the company's data that are associated with the particular called party identity *general manager*, regardless of who the person actually is. Again, the personal function will be independent of the terminal or the NAP the person uses.

   People can have more than one personal function. For example, the individual in our example could be both general manager of company ABC and chairman of the local golf club.

3. *The terminal identity.*
   In future, many different types of terminals, with different technical capabilities, will be available on networks. Consequently, it is extremely important that users be able to have their terminal identities transferred when they call another terminal or a service. Perhaps, for example, a user's terminal can only be used locally in a particular area. Or, perhaps the terminal has technical limitations that prevent it from connecting to a called party. The terminal might, for example, be a normal phone, and the user is trying to reach a service only created for broadband access.

   This terminal identity will be independent of who uses the terminal or from which NAP the calls come.

4. *The terminal function.*
   It might be that a terminal can only be used by a certain group, for example, only the members of the local golf club and no one else. A terminal can also be dedicated for emergency calls only, with geographical restrictions, and so on. Again, this terminal function will be independent of who uses the terminal or from which NAP the calls come.

5. *The physical location.*
   The calling identity might be defined by the physical location on the network, for example, the NAP. If someone wants to order a pizza or call for assistance when

his or her car breaks down on the highway, physical (geographical) location i
what is important. And physical location will be independent of the person wh
calls or the terminal used.

Furthermore, when parties make or receive calls, they will have to distinguish be
tween their different roles in some way. This can be done in several ways, depending o
the type of mobility implemented. Identities can be distinguished:

- Through the use of identification cards (see Section 6.1, item 8). By inserting differ
  ent cards into a terminal, users can distinguish between their possible roles as privat
  person (calling person) or personal function. These cards can also identify the differ
  ent functions of a certain terminal and can be used in combination with a PIN cod
  for security reasons.
- Through the use of different access codes, that is, different numbers for private per
  sons and individual personal functions.

Combinations of the five identities described in the preceding list represent the nex
step. For example:

- The general manager (from item 2) may only be able to make the type of call h
  wants (for example, international) if he is using a certain type of terminal (as in iten
  4) in the company (without geographical restrictions).
- A person (item 1) maybe only get full access to special information if she is locatec
  at a special location, for example in her office (as in 5).
- A person holding a function (item 2) and working partly at home may have all hi
  calls paid by his company if he is calling from his home (as in 5).

A next step is to combine these five types of calling identifications with time de
pendence:

- The person working partly at home might only get his calls paid by the company i
  he calls during working hours.
- A person might be identified by her function when she calls during working hour
  and as a private person during her leisure time without having to change these role
  manually.

The Called Party

Like the calling party, the called party can be structured according to personal identity
personal function, terminal identity, terminal function, or NAP.

When you call someone today, you usually need to know the exact location (th
NAP) of the person and the number to that location. Furthermore, you must remembe
or look up the number you dial. Now, in about 50% of the cases, there is no answer

Value-added services like call forwarding and others can increase this figure somewhat but, based on experience, not much. This is because people forget or do not bother to use these services, mainly because of bad user interfaces or poor knowledge about service benefits.

The ultimate goals for calls in future should be to have much better terminals than are generally available today and to use letters instead of numbers when dialing. We would also like to have a more *personal* way to call, for example, when I dial "dentist" I want to reach *my dentist* (a personal function) and no one else. In the same way, my neighbor wants to reach his dentist (not mine) when he dials "dentist." But, on the other hand, when we dial "plumber," we might want to reach the same person. So, if we could find an easier way to dial or, perhaps, just speak to the terminal instead of dialing, these would be tremendous improvements over communication today.

Call Forwarding Services

The first service in this category that comes to mind is call forwarding unconditional (CFU). We must, however, also consider call forwarding in combination with a condition, like no reply or when busy. Consider the following example.

Mr. Smith is the general manager of company ABC. He is also chairman of the local golf club. Mr. Smith is married and lives in a house near his work and near his golf club. For his different roles, he has different numbers:

- As a private person, his number is 111 (a personal identity);
- As the general manager of company ABC, his number is 222 (a personal function);
- As chairman of the local golf club, his number is 333 (a personal function).

At a certain time, perhaps calls to all three numbers are routed to his car terminal. For example, let us assume that he is on his way to the airport in half an hour and he wants to be sure that he does not miss any calls while he is abroad, as he cannot answer any of them outside the country.

He cannot use call forwarding to forward all three to only one person. As a private person, he probably wants his calls to go to his wife, that is, he initiates call forwarding unconditional (CFU) from a personal identity to a personal identity. As the general manager, he wants his calls to go to his secretary, initiating CFU from one personal function to another personal function. As chairman of the golf club, he has four choices: (1) the secretary of the golf club (CFU from a personal function to a personal function); (2) the clubhouse (CFU from a personal function to a physical location); (3) the mobile terminal owned by the golf club (CFU from a personal function to a terminal function); (4) Mr. Haley, a friend and a fellow member of the golf club, but with no official function there (CFU from a personal function to a personal identity). Obviously, Mr. Smith will have a lot to do before his plane leaves in half an hour.

Let us assume that Mr. Smith succeeded in initiating all the CFUs before he left. When he returns once week later, he will have to transfer (collect) all his numbers again. He will have to remember where all his CFUs were made and, after talking to everyone and canceling all the CFUs, he will have a clear picture of what happened at his home, at his company, and at the golf club while he was away.

In this example Mr. Smith only had three roles. In future, people could have perhaps five or more roles. It might then be very easy for them to forget where they have assigned calls for each of their roles while away. All CFUs could probably be canceled, and the roles (all of them) could be taken back immediately, so this is not a problem. However, if they cannot remember who received their calls in the separate roles, these people will have a problem. How will they find out what happened during their stay abroad? Obviously, we will need some type of function that provides a complete list of all our numbers (for all our roles) and where they go at a particular moment in time. We will also probably need to combine this function with a history function, so that we can find out what happened yesterday, for example.

Concepts Such as Virtual Private Networks

Full mobility is very well suited to virtual private networks (VPNs), especially when full mobility is being created inside a VPN. In general, the preceding discussions about calling line identification and call forwarding for networks is also valid for mobility inside a VPN.

Services Independent of Full Mobility

Some services will not be affected by the introduction of full mobility and can be used in fully mobile networks. Such services are often of the type *call initiating* and include:

- Abbreviated dialing, which translates a short number into a real one for initiating a call;
- Repetition of last dialed number, which repeats a number for initiating a call;
- Wakeup service, which initiates a wakeup to a predetermined number;
- Hot line, which initiates a call to a predetermined number.

These services also make good candidates for implementation on terminals, as we have seen in previous chapters.

6.3 SERVICE ADMINISTRATION AND MANAGEMENT EVOLUTION

The area of service administration and management is a key area that will impact the future evolution of service implementation. This area will be important because it will influence to what extent we can

- Expand the number of services and customized services on networks;
- Allow end users (service providers and subscribers) access to data and program areas in SCPs and SDPs (either directly or indirectly via SMP);

- Reduce the time-to-market period for new basic services, service provisioning, and customization of services.

One of the difficulties of describing systems for service administration and management for a network is that they are mostly either vendor-dependent or network-dependent or both. This means that they are frequently different in different networks and greatly influenced by the specific products for intelligent networks, switches, and other systems on the network.

Administration and management systems must, however, follow the same stringent rules of flexibility as technical systems must in the introduction of an intelligent network. This means that the ability to rapidly introduce a new IN-based service depends on fast and easy updating of the entire chain of systems involved, including technical and administrative systems. The latter include systems for customer databases, service administration, charging, and billing. To achieve this flexibility, we must also use an IN-based approach in building these systems. This forces us to use a platform concept whereby administration and management of new services, service provisioning, or customization of services can be introduced easily and quickly.

When a service is to be closed for some reason, it is perhaps even more important to be able to close all involved systems quickly. Perhaps the service is to be closed to a certain user or completely withdrawn from the network. The reason why it must be done extremely rapidly is often related to such matters as misuse of a service, fraud and abuse or the risk of these, or the service's interaction with another service

A service administration and management system is an all-important item on a network, concerning almost everything around the service. For this reason, matters pertaining to administration and management of services are also described in other parts in this book. This section is not intended to present a comprehensive description of future administration and management systems, but will instead discuss some of the most important evolutionary steps in the future of the intelligent network.

The remainder of this section is divided into discussions of future:

- Service prototyping, testing, and supervision (Section 6.3.1);
- Evolution of customer control capabilities (Section 6.3.2);
- Charging and billing systems (Section 6.3.3);
- Security and reliability demands (Section 6.3.4).

6.3.1 Service Prototyping, Testing, and Supervision

Introduction of the intelligent network marks the start of a more dynamic offering of services from network providers. Intelligent network technology allows much faster creation, provisioning, and customization of services. Customized services—either single or whole service packages—can be provided either to single users or to groups of users, such as pizzerias, banks, insurance companies, and households.

Many services will be here to stay, while others will only be temporary. The large number of services we will soon see brings up an important issue for the network operator. Operators will need an efficient cleanup facility to take care of unwanted services or, in other words, a simple way of removing services. This is very important from an administrative viewpoint because the more services exist on a network, the greater the risks of mistakes and of unexpected behavior, such as in service interaction that is not anticipated.

Even with a cleanup facility, a much higher number of services will be available on all networks in future. This fact, together with a shorter time-to-market period for new services, places a greater demand on the need for extremely good service testing facilities. Types of service testing that will be required include:

- Testing before a new service is introduced;
- Testing of service problems that appear;
- Prototyping, that is, testing different technical solutions or testing new services on the market;
- Testing services for interworking purposes.

To fulfill these testing requirements, tools must be developed. The types of tools are described in the sections that follow.

Tools for Prototyping

In future, new trends will appear in society and new demands from different business areas will arise. These, in turn, will require networks to provide new services. It is very important for the networks to be able to react quickly and introduce a required service rapidly. We have already discussed this and, as we have seen, the intelligent network can accomplish this.

But being able to react quickly does not mean following all new demands blindly. What we must be prepared to do instead is rapidly evaluate whether a service demanded should be introduced or not. And, if the answer is yes, we must be able to introduce it quickly. Also, even though we can do it quickly, we must always select the right time for introduction, because it might be better to wait and see. Timing is important!

Following this consideration, we must choose the appropriate method of implementation, as we have discussed previously. Should the service be implemented on the intelligent network platform, on the nodes, or on the terminals?

An interesting possibility that intelligent networks will give us in future is a tool for providing services on a *speculative basis*. Network providers can use this tool for services they are not sure they want to introduce permanently on the network, for example, because of risks of abuse or fraud or because they are not sure what the market wants. The tool allows a network operator to quickly introduce a service maybe only in a geographically limited area in order to monitor its performance. If the service turns out to be a problem, it can be easily closed again. If it performs well, it can be introduced permanently network-wide.

Tools for Testing Services Before Introduction

In addition to normal software tests, there are other tests that must be carried out prior to network introduction. The increasing number of services, customized services, and service dialects calls for testing of the interworking of a new service with existing services. Of course, a service interaction test is the most urgent requirement; clearly, testing for negative service interaction is highly important in this context. (Please see Section 6.4, which deals exclusively with service interworking.) Eventually, all networks will need to possess a tool that shows the effects of interworking between existing services. New services can be tested by this tool before they are added to the network.

Tools for Number Changes Made on a Network

The fact that the service control points (SCPs) used for number translating in IN-based services can also be used for temporary number changes by the network operator is often overlooked.

One example involves the necessity of changing the number for a service. Number changes are often hard to carry out because they involve changing the habits of many customers. A number frequently used to reach a service, a function, or a person is not so easily forgotten. Moreover, it is not easy to remember a new number. In addition, old numbers are often heavily advertised in telephone directories and newspapers and written down in millions of private telephone books. So, when a new number is opened and an old number is closed, the old number will still receive a great deal of calls before people manage to learn the new number. Experience shows that calls to old numbers can continue for several years.

In this instance, the SCP on an intelligent network can be employed as a tool in the following way. When the number change has been made, all exchanges on the network are taught to allow only calls made to the new number to reach the service. When someone calls the old number, the call is transmitted from the local exchange to a service switching point (SSP), which calls the SCP. The SCP translates the old number to the new one and sends it back to the SSP, which sends it back to the local exchange. Now, the local exchange performs the service, but also gives a recorded message stating the number is closed and the following number should be used in future. The SCP provides this translation service for a certain period (a year or so). After that, the old number is treated like any other vacant number. Using the SCP instead of the local exchange allows the function to be centralized. Everyone receives the same message at the same time, and if the message is changed, it is changed simultaneously throughout the whole network.

Tools for Following Up on Service Problems

In future, all new services, customized services, and service dialects will probably cause users to make many mistakes. It could be very difficult to follow up later and correct these mistakes unless we can save a record of the actions carried out by the users. In addition, when abuse and fraud occur or are suspected, it is highly valuable to monitor the services involved. Those two examples point to the need for a *logging*, or *history*, function for services. Logging functions are usually included in the software provided with an intelligent network installation. (Automatic message accounting (AMA) technology, which is primarily used for charging purposes, could be of use in some special cases of logging. See Section 6.3.3.)

A logging, or history, function allows us to go back and see what has happened in the event of a problem and to print out a list of the last activities carried out by the users. Different countries have different laws about this kind of supervision, which must, of course, be considered.

6.3.2 The Evolution of Customer Control Capabilities

The first years of customer control on the intelligent network will be characterized by users having the capability to make simple changes in their own data, for example, changing routing times for number translating services or changing passwords for credit call services. Further evolution in this area will probably be determined by the availability of user friendly terminals that offer guidance in the use of services but also of the availability of good security systems. In other words, evolution will depend on the quality of the user interface. Security is the most important aspect of customer control because a network operator's worst nightmare is to have hackers enter the system via the customer interface.

Different countries are experimenting with introducing a voice-controlled customer interface, that is, allowing users to speak instead of using the buttons on their terminals. The voice is checked and read by voice recognition equipment and translated into words. There are, however, problems with this type of use and we will probably have to wait some years before it becomes commercially operational.

The main problem remaining to be solved involves the system's ability to understand the word that has been spoken and not confuse it with another word. Consequently, the number of words used in systems today (1994) is often very limited. Usually, not more than 20–30 words, including the numbers 0–9 and "yes" and "no" are available. And even with this limited set of words, the technology can not promise a 100% probability of hitting the right word. Moreover, people pronounce words differently, use more words than they should, cough or make other noises, and so on.

Security poses another and probably more severe problem for voice-controlled identification services. These services would involve assessing whether, for example, the individual trying to make a credit call or change the routing of a freephone number is the right person. Even if we succeed in distinguishing between different voices, we can never be sure whether we are hearing the person speaking or a recording of the person.

Later in this section, we look at the future evolution of how users (customers) and the network will interwork. While possible solutions will mainly be based on the intelligent network platform, nonintelligent network solutions may also offer interworking opportunities on some networks.

A Dynamic Customer Control Function

The customer control function—permitting customers to enter their own data directly into the SMP by using a password from their terminals—will be used in a more sophisticated way in future. The first type of use can be described as *static*. Static uses will include changing answering locations, times, and passwords for employers. For an example, let us return to the pizzeria described in Section 4.3.3. The general manager changed the distribution from 25-25-25-25% to 28-28-28-16%, because of a temporary shortage of staff at one pizzeria. But even if the general manager is able to change the distribution between the four pizzerias, there will still be a period between the time the problem occurs and the time the effects of the change are noticed. Perhaps the change was not adequate. There could still be some variation in the load between the local pizzerias that could prompt another change. During this period, which could run into several hours, many customers could be lost.

Of course, this situation is far better than if the pizza company had no system for load regulation at all. But if we could create a more *dynamic* system that changed the distribution between the local pizzerias more rapidly when the need arose, we would have a more flexible and more customer friendly system. What are the main parameters in this example? They are

- The delivery time, from receipt of an order until the pizza is delivered;
- Creating a continuously even distribution of delivery time between local pizzerias.

The delivery time depends on such factors as

- The number of orders and their distribution over time,
- The number of pizza chefs available,
- The number of delivery drivers,
- The number of vehicles available.

To simplify matters, let us only consider the delivery times for pizzas in each location. Let us also consider only a region of four pizzerias and a normal distribution of 25-25-25-25%. Each local manager is asked to keep an eye on the time from receipt of an order until the pizza is delivered (this is probably their normal job). The local managers call in this delivery time to the SMP using their advanced customer control function. The reporting should occur once an hour or sooner, if necessary, as when, for example, a drastic change takes place.

The SMP makes a calculation based on the times reported by the local managers to update distribution between the local pizzerias. The SMP then sends this information to the SCP for execution. Suppose the delivery time is 15-30-30-50 minutes for the four pizzerias. The fourth pizzeria has a considerably longer delivery time than the first pizzeria. When all local managers have reported their figures, the SMP calculates a change from 25-25-25-25% to a new distribution of 35-25-25-15%, which is then sent to the SCP. The next report recommends 40-30-30-20 minutes. The change was obviously too large. SMP then changes the distribution to 30-25-25-20%. The SMP keeps an eye on the delivery time and tries to keep it as balanced as possible.

This dynamic system continuously strives to maintain the same delivery time for all four pizzerias. In addition, the working hours of the pizza chefs could be sent in by management, if they vary. If the SMP knows that at 5 pm the staff in one location drops from four to two chefs, it could automatically change the distribution at that time.

The purpose of this simple example is to provide an idea of the capabilities that intelligent networks of the future offer to businesses. The assistance described could be extended even further by taking into account not only the number of pizza chefs, but the number of drivers, the number of vehicles, and so on. Furthermore, the cash registers at the pizzerias could be directly connected via a telephone line to the SMP. The SMP could then automatically monitor delivery times from the moment an order is called in until the cash register prints the bill and the pizza goes out the door. In this way, the SMP could be an even more sensible and dynamic instrument for load sharing between the pizzerias, performing a continuous change of the distribution.

Reading Out Your Own Data

Another use that will increase in future is the ability of users to read their own data from the network before making changes. Networks become more complex as users become increasingly mobile, and they contain more and more services with such features as time- and origin-dependent routing and password requirements. It must therefore be possible for users to employ passwords to read their personal profiles, either at their terminals or via facsimiles or the like.

Another example is the ability to read out the value of your own call meter, that is the actual value of charges. This service has been introduced in Sweden and other countries and is used in various ways. For example, if you rent out a holiday cottage, you can take readings when your guests move in and when they move out and calculate how much they should pay for the use of your phone. A customer control interface to the AMA system could also provide you with this facility (see Section 6.3.3).

A third example that will probably be highly valuable in future is the ability to receive a copy of all services activated at the moment or a copy of your service profile.

Third-Party Service Creation

Once it becomes possible to build intelligent network services by combining service-independent building blocks (SIBs), which can be accomplished in weeks or months, the next question is, can this building be performed by parties other than network operators? Perhaps, in the future, users (customers) or hired consultants will be able to use the service creation tools provided by network operators to build their own services. But whether or not this can become a reality does not depend on the knowledge of third-party service creators. Service creation can be learned and is fairly easy. The most difficult part involve security. A service created by a third party must of course be tested for software quality, but that is not enough. Testing of service interworking is also essential. If, when it is introduced on the network, a service causes interaction problems with existing (or anticipated) services, we should very soon run into great problems on all networks.

Network operators who offer third-party service creation capabilities must therefore have *extremely good tools* for testing services before network introduction, both the software itself and the service's interaction with other services. My personal feeling about third-party service creation is that if it does become available in a broader scale, it will not be in this decade due to the security aspects.

6.3.3 Charging and Billing

A key area in which we will probably see one of the greatest revolutions on most networks in the next 2–5 years is charging and billing. In this period, we will see a large number of new services, the introduction of greater mobility within and between networks, and increasing competition in the telecommunications area, to name a few trends. These trends will, in turn, fuel the need for more flexible and reliable charging and billing systems.

Of all the systems involved in service provisioning, none differs more between networks than charging and billing. Some are based on a modern AMA record technology (also called toll ticketing), while others are based on the old and inflexible call metering (CM) technology. Increasing use of intelligent networks will no doubt speed up the introduction of AMA record technology in networks, at the expense of CM technology.

The next sections describe both CM and the AMA record technology. The section on AMA technology explains how automatic message accounting can be used on intelligent networks, how it can be further developed, and what its benefits are.

Call Metering Technology

CM technology, which is described in Section 3.6.2, is the old, basic method of charging and is built on the principle that subscribers each have their own call meters, often located at their local exchange. The call meter records increments in a predetermined way for each call or service used. For a local call, the meter might record a single increment or it might record continuously in a low, frequent manner. On the other hand, for an international call, it will record increments much more frequently. Consequently, different charges are levied depending on the call or the service used. The call meter is read regularly, the value is compared with the last reading, and the difference forms the basis for charging the subscriber. The reading, however, provides a value only. Moreover, it is impossible to distinguish local calls from long distance calls and other services. And you cannot provide your customer with value-added facilities like specified (detailed) bills, and so on.

Call meter technology fulfilled the requirements for charging as long as the network was only used for basic telephone calls. The situation today, however, is different. Services are more heavily used, and the telephone network is becoming more important for business users. Competition, both between different network operators and between telephone networks and other media, such as newspapers, television, and radio, is also increasing. These changes call for a more flexible system in which each activity can be sorted out according to type, time of occurrence, and duration. The answer is the introduction of AMA record technology or toll ticketing (which is the name used primarily in Europe).

Automatic Message Accounting Record Technology

Automatic message accounting (AMA) record, or toll ticketing, technology (also described in Section 3.6.2) shifts charging and billing from the network to an offline computer system, which receives input from the network. With this technology, information is logged at one place, the local exchange, on the SSP or the SCP, about

- The calling party;
- The called party;
- Other parties involved, if relevant;
- The activity (for example, a normal call, a service, a failed call attempt);
- The time and date of the activity;
- The duration of the activity.

By activity, we mean anything that you can do from your terminal, for example, make a normal call, use a service, use a customer control function on the SMP, or simply lift your handset. The information about the activity is transferred to a separate system, not on the network, where recording and billing take place.

Having all charging and billing performed off the network allows the network to become much more flexible than it was when CM technology was employed. Benefits obtained from an offline AMA record system include the capabilities to:

- Charge any party, including both parties (split charging), a third person, or an account;
- Charge for activities other than normal calls, such as for use of the customer control function on the SMP;
- Possess more flexibility in charging, for example, charging different users differently or charging different rates at different times;
- Obtain statistics from the network about when calls were made, their duration, from and to which region they were placed, and so on;
- Output failed calls;
- Specify all or some of a customers' calls on their bills;
- Cooperate with other network operators and service providers for premium rate services, since all activities traversing interfaces are logged.

This information will be easy to output with the help of AMA record technology. If customer control is added to the AMA record system, we might also receive

- Immediate information at terminals about the cost of the activity just carried out;
- General information about the cost of carrying out a certain activity, such as making a call from A to B at a certain time and date;
- Information about the total cost of subscription from a certain date;
- The ability to change the charged party during or after a call.

These on-line functions might also be offered by advice of charge services, with or without the assistance of the AMA record technique.

For IN-based services, it is often preferable to have logging of charging information performed by the SSPs.

There are some functions easily provided by CM technology that are also available from AMA record technology, with perhaps a few implementation changes. Examples include services for limitation purposes, often ordered by the customers themselves, such as

- Time limiters, which only allow calls to last a certain length of time;
- Geographical limiters, which only allow calls to be made to a certain area or areas or received from a certain area or areas if freephone service is used;
- Limitations on expensive calls, such as international calls and calls to premium rate numbers.

The Future Points to a Charging System More Dependent on the Value of a Call than the Distance

Imagine, for instance, that you want to call your representative at your insurance company to discuss your car insurance. But your insurance company made sudden organizational changes and your representative has moved 1000 miles away. The representative still gives you exactly the same service as before, but you are probably not willing to pay more (pay for a longer call distance) for the same service.

Historically, the charging of a call has been dependent on the distance between the two parties. A call that needs to travel a longer route on the network (and uses more network resources) has always cost more than a call that uses fewer resources. This is the basic philosophy for charging. It is also been based on the network operator's cost for the call, 80–90% of which are network transmission costs. The remaining 10–20% are the switching costs (costs for local and transit exchanges, etc.).

Although this form of charging is expected to remain dominant for many years, there is a movement towards a more distance-independent charging based on the benefits of the call or the service to the customer. In fact, a number of trends point in this direction:

- A continuous decrease in the cost of bandwidth, that is, for transmission, due to a greater use of high bandwidth connections, fiber optics, and radio-based transmission. This will drastically change the relation between transmission and switching in favor of the former.

- A continuous increase in the cost of "intelligence," that is, for the intelligent network, used for number translating purposes, customer control, and so on.
- Increasing competition between network operators.
- Increasing use of distance-independent services for the calling line, such as services like freephone (distance-independent for the calling party), televoting, premium rate, and so on.
- Competition between services, such as premium rate and other media outside the telephone networks.
- A continuously increasing share of total revenues for the network operators derived from IN-based services that are distance-independent.

6.3.4 Security and Reliability Demands

Passwords are the security tools that normally come to mind in discussions about security. Passwords are, in fact, a good tool for preventing abuse and fraud.

In discussions about reliability, issues such as MTBF (mean time between failures) and MTR (Mean time to repair) for terminals, switches, and so on come to mind. There are, however, other things we can do to increase the level of security and reliability on a network. One way is to use some of the new services, which are often IN-based, to raise security and reliability levels for network operators, service providers, and subscribers. Examples of services that can be used in this way include:

- *Identification services*. Calling line identification and called line identification facilities are good ways of identifying who you are talking to. They do not provide a 100% guarantee, however. (These are discussed in more detail in Section 6.2.2.) The use of cards and of account numbers in combination with passwords are other security facilities.
- *Virtual private networks*. These provide security because of the special routines required for calling inside the VPN, compared with making calls to or from the VPN. The security lies in the knowledge that you are speaking with someone from inside your own VPN (that is, inside your own company).
- *Customized services*. Customization can increase security and reliability, fitting services to the special security needs of a customer.

Other ways to increase the level of security and reliability include:

- Customer premises equipment (CPE), or equipment residing within the user's domain. If the equipment is located in your own physical domain (inside your building), you have total control over it and thus a high degree of security.
- Priority, where high priority can be used to ensure a fast and reliable (no congestion) connection through the network.
- Low bit error rate routes on the network.

- Testing, in the SCP, of the two parties' access forms. For example, if one PSTN access tries to call a broadband access, the SCP will forbid a connection or guarantee a connection according to the conditions of the weakest party.

Security and reliability also offer the possibility of differentiation. Different levels of security may be offered to different users (subscribers) depending upon demand and cost.

6.4 SERVICE INTERWORKING

Service interworking refers to two or more services used together during a call, either simultaneously or in sequence. A call involving two services is the most common, however three or more services are possible as well. These services could all be initiated at the calling party, at the called party, at a third party, or at a combination of parties. Although service interworking usually causes no problem, in some cases the result is not what we expected.

6.4.1 Understanding the Complexity of Service Interworking

As the number of services and service variations increases with intelligent network use, interest in service interworking also increases. Testing all possible combinations of interworking between services, two by two, was a complicated task even before the introduction of intelligent networks, when a normal network often contained no more than 20 services. It was complicated because every service needed to be tested against all the others, and the number of combinations to test is of the magnitude of the square of the number of services divided by 2, that is, $\frac{20 \times 20}{2} = 200$, considering two services at a time of a possible 20. When intelligent networks are introduced, the number of services to consider and test will soon reach 100–500 or more. And in addition to this, each service *variation* (customized service) must be tested as well against all the others. Clearly, the number of test combinations will soon be too many for conventional (manual) methods.

To further complicate the situation, we must remember that, so far, we have only discussed a situation in which all the services are on a single network. To ensure complete control of service interworking for subscribers, we must also consider services that link subscribers from different networks. Here, both subscribers may have a different set of services and may also belong to different network operators.

As if this were not enough, there is still another dimension to consider, namely, the number of *service dialects*, that is, different methods of implementation for one service, that exist. Two services can behave towards users (subscribers) in exactly the same manner, but if they are implemented by different methods then interworking with other services could cause them to behave in unexpected ways.

And finally, as was previously mentioned, we must also consider situations in which more than two services work together.

So, testing every conceivable service interworking scenario that subscribers on a network might experience is quite impossible—an enormous number of combinations could occur. Instead, we must find a more pragmatic way to treat this problem. We discuss this solution in Section 6.4.4.

6.4.2 Structuring Service Interworking

Before we try to treat negative service interaction, we must find a way to structure service interworking in general, so that we can identify the interesting parts and ignore the uninteresting parts. For the purposes of our discussion, we will formulate a structure for a scenario in which two services are used together. Let us consider two services, S1 and S2. Based on the definitions of the services, the following interworking scenarios are possible (see Figure 6.7):

(a) S1 and S2 cannot be used together because they are alternatives to one another or because there is no reason to use them together (Figure 6.7(a)). Examples include:
- Call waiting and call completion to busy subscriber,
- Call transfer when busy and call waiting.

(b) S1 and S2 can be used together because the function of S1 and S2 together is as expected, that is, the function is exactly the sum of the two single services (Figure 6.7(b)). Examples include:
- Abbreviated dialing followed by call completion to busy subscriber.
- Credit call and remote control of call forwarding unconditional, forming together a simple personal communication system (PCS). (See Section 4.2.1.)

(c) S1 and S2 can be used together but the function of S1 and S2 together is not exactly the sum of the two single services, that is, it is not what was expected (Figure 6.7(c)). This is what normally is called service interaction.

Service interaction will most probably yield a negative result to users, but it may be harmless. In some rare cases the result might even be positive. Effects of service interaction may differ, but usually

- One service overrides the other, that is, only one service is executed;
- There is a fault in charging.

Examples of situations in which service interaction might occur are in

- Automatic alarm service and call forwarding services,
- Consecutive call forwarding (CF).

These are described in Section 6.4.3.

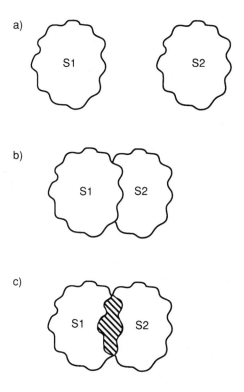

Figure 6.7 Service interworking possibilities: a) mutually exclusive services, b) combined services, c) interacting services.

The same interworking scenarios, with the exception of scenario (a), which can never occur, are valid when three or more services are used together. In scenario (b), the services are behaving as expected and the results do not need further explanation. In scenario (c), however, service interaction could cause problems if we are not careful. And the problems will increase as more services, customized services, service dialects, and communications between different networks become involved, which, unfortunately, is exactly the goal for the intelligent networks.

6.4.3 Service Interaction

What makes the question of service interaction really complex is, as we have seen, the great number of customized services and service dialects and all the possible combinations of calling and called parties that can occur on different networks. If we compare different types of service interaction, that is, compare interaction at the specification level with interaction between customized services, and finally with interaction between service dialects, the latter are the most complex. Their impact on the network and on other services is often more difficult to estimate for these reasons:

- Service interaction at the service specification level is, of course, found in specifications. The service can be a standardized service or a proprietary one. In either case, the specification can be read and understood and so interaction with other services can be anticipated.
- Specifications for a customized service will also indicate whether interaction with other services can be anticipated.
- Dialects of services, however, are dependent on the particular implementation method employed, and the implementation specifications are not always easy to find (particularly if the service was implemented on another network). If they are found, they are not always easy to understand.

We will look at some examples of service interaction on existing networks, that is, networks on which mobility (Section 6.2) has not been introduced. The examples were chosen to illustrate different types of service interaction. Following each example are comments about the effects that occur.

Automatic alarm service and call forwarding, no answer. These two services are not normally customized today, but different service dialects may occur. The specifications of the two must be regarded as clear. The automatic alarm service can be used to wake you up in the morning. You initiate it from your phone and dial the time at which you wish to be awakened. At the wakeup time, your phone rings. Call forwarding, no answer works as follows: If someone calls you and you do not answer within a certain time, the call is forwarded to another number, which you dialed in when you initiated the service.

The two services used together pose no technical problems for the network, but there is one unexpected *social* effect: If there are no restrictions on using the two services together, you can initiate call forwarding, no answer and forward the call (the wakeup) to someone else, that is, play a practical joke on someone. This has been foreseen in most networks, however, and the combination is normally not permitted. Instead, a wakeup call must always terminate at the phone that activated it and may never be forwarded to another number.

As we can see, these two services had an undesired social effect that forces network operators to forbid their combination.

Consecutive call forwarding services. These services usually work without any problems. But we could easily end up with endless loops if we are not careful. For example, A makes a CF to B, B makes a CF to C, and C makes a CF to A. This is another reason why network operators should restrict the use of some services. One of the most common restrictions is to forbid certain combinations of services that *tend to provide the risk of being too complex.* The solution for avoiding loops in the network is often to allow only two consecutive call forwardings to be executed on one call. Restrictions are placed on call hold in combination with call transfer for the same reason.

These services are an excellent example of when an IN-based implementation in a centralized SCP is preferable to a node-based implementation, as in future the SCP can have an overview of the activated services on a network and thus can avoid setting up connections that cause loops.

Call hold and call forwarding unconditional. Interaction can also be dependent on implementation, as in this example and the example that follows, which were both found (and solved) on the Swedish public switched telephone network (PSTN). Call hold (CH) works as follows: A calls B and during the call B initiates a CH and leaves A temporarily to make another call. During or after this second call, B may go back to the call with A again. Call forwarding unconditional (CFU) works as follows: Once you have initiated it on your phone and entered another number, all incoming calls will go to the new number until you change it. In the Swedish PSTN, CH and CFU work nationwide. This scenario illustrates what used to happen when the two services were used together.

John calls Mary. After a conversation, they decide to go to a movie in the evening. Mary suggests that they ask Elizabeth to join them. While she is still on the line with John, Mary initiates a CH and places a call to Elizabeth. But Elizabeth has initiated a CFU to Brian. What would happen, however, was that the call initiated after the CH would break through the CFU. Mary's call was supposed to go to Brian's phone (because of the CFU initiated by Elizabeth). Instead the call would go to Elizabeth's phone (where no one was at home).

This interaction was *implementation-dependent* and has been solved on the network.

Freephone service and call hold. Another scenario involving call hold illustrates implementation-dependent interaction.

The cinema in town has a freephone number for ordering tickets. Mary calls John and after a conversation they decide to go to a movie in the evening. John asks Mary to hold for a moment while he initiates a CH and calls the freephone number to order tickets. However, it was impossible to do this on the Swedish network in the beginning of the 90s. The reason is not so easy to understand, but it was due to the implementation (service dialect problem) of the freephone service and the charging system that was used during the first years of intelligent networks in Sweden. At that time, the AMA technique for charging had not yet been implemented. The call metering pulse technique was used to calculate the charge for the owner of the freephone number via a complicated procedure. Actually, the SCP translated the freephone number to an intermediate number series. This intermediate number then used a CFU (hidden from John) to the C subscriber. So when John tried to call the freephone number, his attempt was rejected because it was initiated after a CH.

Because of the interaction of CFU and CH, there was also an interaction between freephone and CH, because of the hidden CFU used by the freephone service. The implementation of freephone has now been changed and this interaction has disappeared.

6.4.4 Solutions to the Service Interaction Problem

As we have seen in the preceding sections of this chapter, the overall service interaction problem is enormous. There is no single solution to the problem. Moreover, there is today a lot of hidden, undetected interaction between services on all networks—between services within a network and services that interact across networks. Most interaction cases are harmless or cause very little damage. However, a few cases could be really severe and might cause major problems.

Considerable efforts are being devoted today to trying to solve the more general service interaction problem. Nevertheless, because of the complexity of the situation, it will probably be a long time before something is developed that can help us on the networks.

What can be done in the short term and what also might be the best way to avoid the worst cases is to use a more pragmatic approach, identified in the following guidelines.

(a) *Identify the most probable instances of severe service interactions.*
First, identify the combinations of services that might cause trouble. Then, determine the probability of each combination occurring. Use those two parameters to determine what to focus efforts on first. At the third stage, consider the severity of the combinations, that is, the damage the combinations will do. The probability that a combination will occur is, of course, greater if both subscribers are on the same network, for example, on the Swedish network, than if one is in Sweden and the other in Australia. Put more effort into solving the interaction of high-probability combinations, even if the problems they cause are relatively small. Rare combinations can be left for the future or at least be considered lower priority.

(b) *Conduct investigations on the networks.*
If you decide that the negative service interaction between two subscribers is the most serious problem, you can start your investigation here. This is best done by setting up a matrix with all services on both axes. Remember to include customized services and service dialects as well. Then, go through all combinations to find the important ones, which is a time-consuming task. You will soon find, however, that many combinations of services can be quickly disregarded because they cannot be used together, either because they are used in different phases of a call or because they are alternatives to one another. The remaining combinations must be checked carefully, however, to determine the possible risk of negative service interaction.

Once you have a complete list of all service interaction cases, the next step is to try to solve the problem. However, many of the interactions are not so easily solved. Some are, but not all. So, what you can do instead is follow the procedure described in (c).

(c) *Put up a warning flag of the risk of certain combinations.*
When customers gets a service package with a subset of all the services available on the network, they can be warned that a certain service in certain situations may cause problems.

(d) *Test the interaction between different customers' service packages.*
If customers call other customers, who possess different service packages, the network can warn them about the risks of using certain combinations of services.

(e) *Create a tool to test new services.*
A tool can be created based on network service interaction. When a new service is to be introduced, it may be tested with this tool to find interaction risks.

(f) *Forbid certain combinations, when necessary.*
High-risk combinations might be forbidden. This might often be carried out in a manner whereby one service has priority over the other. In this case, when you try to use the second service, nothing happens, you may only get a message that the service cannot be used in this situation.

So, in the short term, we have to learn more about the service interaction problem. In the long term, which is perhaps 3–5 years, I believe that we must develop *really good tools for testing and avoiding service interaction.* Our failure to achieve this could pose the greatest threat to future expansion of services and network interworking and, consequently, to the entire intelligent network concept.

REFERENCES

[1]   Andreae, Edfelt, Fröberg, *Citatboken*, Stockholm, Natur och Kultur, 1991.

Chapter 7

Future Network Evolution in Harmony with Intelligent Networks

The right *resource* at the right place at the right time.

The ideal goal for continuous development of network platforms in the future, both technical and administrative, must be to fulfill the needs of *all* the users at *all* times. In Chapter 2, users are defined to include such customers as subscribers, service providers, and network operators. The subscriber can be divided in two parts, the service subscriber (who pays) and the service user. The needs of different users are discussed in Section 7.1. Different customers and different types of traffic will place different demands on future networks, as we discuss in Sections 7.2 and 7.3. Section 7.4 covers the impact of these demands on future networks, including how to expand reach for both existing and new services. Network economy is always an urgent matter for network operators. Consequently, I cover certain aspects, such as resource allocation and resource optimization in the networks in Section 7.5. Finally in Section 7.6, we look at the possibility of future interoperability of intelligent network platforms.

7.1 DIFFERENT USER NEEDS

In Section 2.5, we identify the users of intelligent networks as subscribers, service providers, and network operators. The subscribers are divided into service users and service subscribers. ("Service" includes basic calls.) The specific needs of individual users differ tremendously, and we examine these needs one by one, bearing in mind that a user can play more than one role, for example, a user can be both a subscriber and a service provider.

7.1.1 Service Subscribers and Service Users

Service subscribers require good response from a network operator, particularly when a subscription is opened, changed, or closed. During the subscription, they demand items, such as specified billing, and so on.

Service users, on the other hand, only see a user interface, which is the terminal from which they access the network and services. The terminal is the only means of accessing the network. When they use the interface, they should, at all times, have the exact level of service determined by their subscription (what they are paying for). No less, no more. Users want to think of the network as a "black box." Use of the network and of services must be uninterrupted, that is, they must be able to have access to the network and services 24 hours a day. Users are not interested in knowing about the technical solutions behind services; they are interested only in the services themselves and in easy methods for using them. On top of that, they want to be able to use the network at any time and regardless of their location.

Figure 7.1 shows how service users in a company use services both locally (in their private domains) and globally (in the public domain).

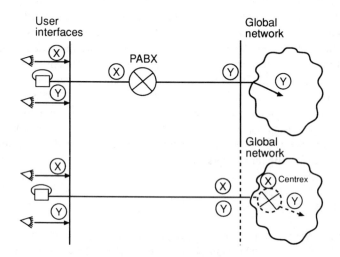

Figure 7.1 Users want to think of the network as a black box.

A user interface consists of both the physical terminal that a user employs and the services that are available from the terminal. Users are connected to a private automatic branch exchange (PABX), which, in turn, is connected to a public network. A service user working, for example, in an office, from the terminal, has an interface to the PABX, which determines the *local* services (including user procedures for local services) he or she can reach on the PABX. We can call this the X interface (Figure 7.1). When the same service user accesses services on the public *global* network, it is through a second interface (but still the same terminal), which we can call the Y interface. This interface determines the available global services, including corresponding user procedures.

When, later on, the company decides to replace the PABX with a Centrex function on the public network, the ideal goal is that service users should not notice the change. They should be able to access the same local services using the same procedures as before, that is, via the X interface. Furthermore, when they use services on the public network, it should be via the Y interface, just as before. If the company accomplishes this, it has created backward compatibility, that is, it has succeeded in carrying out a major change on the network without service users noticing. The next step is to harmonize the user procedures for local and global services, that is, to merge the X interface and the Y interface into a single interface. This is the ultimate goal, as service users would then employ the same procedure regardless of whether they were using a service locally or globally. (For example, users could employ the same procedures for initiating call forwarding unconditional both locally and globally).

7.1.2 Service Providers

Service providers have two roles. They are both sellers to service subscribers and service users and customers for network operators. As customers, service providers have similar demands to those of service users. In many cases, the demands might be even greater, as the network may be the only way the service provider can sell products (for example, audiotex services). Service providers for whom this is true must have 100% access to customers and to the network. Advanced customer control interfaces and terminals, described in Section 6.3.2, are other future demands. In the role of seller, service providers have the same goal as network operators, namely, to provide customers with the services they want at a minimum cost. For example, if announcement machines are used, the location (physical) of the machines, as well as their number, can be optimized to minimize costs. (Please see Section 7.5.)

7.1.3 Network Operators

For our purposes, the term "network operator" refers to both the operator of a network and the operator of services. In future, especially with regard to open network provisioning (ONP), a distinction is made between these roles because they can involve different companies. The ONP concept allows several service operators to work on a platform owned by one network operator.

Network operators are, of course, sellers to service subscribers, service users, and service providers; but they can also be both a buyer and a seller when they are cooperating with other network operators. Network operators are responsible for giving customers the services they have purchased, including the right level of security and availability. Their goal is to do this at a minimum cost while maximizing revenues. Therefore, it is very important to optimize resources and allocate them to the right places on the network. By allocation of resources, we mean both physical equipment (*hard resources*) and software logic (*soft resources*) for service control.

Remember the discussion in Section 2.2.3? "The right *service* at the right place at the right time." There, we were talking only about service logic (soft resources). We now expand this to include physical equipment (hard resources) as well: "The right *resource* at the right place at the right time." We will take a closer look at this in Section 7.5.

7.2 TRAFFIC DEMANDS ON THE NETWORK

Let us take a brief look at six scenarios that illustrate what could happen simultaneously on any given afternoon to calls emanating from the service users of a local exchange:

1. *Fire department call.* A fire suddenly breaks out, and an emergency call is made to the fire department (a voice communication).
2. *Money transfer.* At the same time, the local office of a major bank begins to transfer a large amount of money via the network to a company (a data file transfer communication).
3. *Pizza order.* A call is made from a home to order a pizza for delivery (a voice communication).
4. *Tender facsimile.* A company is faxing a tender to a customer only minutes before the deadline expires (a fax communication).
5. *Computer mail.* A person is sending computer mail via the network from the local office of a company to a colleague in another city (a computer mail communication).
6. *Televoting session.* A television program begins a televoting session (which will result in mass-calling voice communication).

Which of these traffic demands is the most important? Which places the greatest demands on the network? Let us first define what we mean by "important" and "greatest demands." We could mean many things, as we will see, but whatever the case, the network must be able to handle all of these traffic demands in a correct manner.

To start with, how do we know which call is the most important one? Obviously, this is not always easy to say. It depends on our particular frame of reference and is therefore often highly subjective. In these scenarios, we can single out the call to the fire department as the most urgent, as it could involve a matter of life or death. A quick connection without delay through the network to the fire department is extremely important in this context. This is also true during the mass-calling televoting scenario. The transfer of money, in scenario 2, is also important, but places quite another set of demands on the network. Connection within a second is not the primary concern here. Instead, a highly secure connection is the most important demand.

Important traffic always places high demands on the network, but as we have already seen, the demands are not always the same in all cases. If we try to distinguish between different types of *high demands*, we find demands for

a. Real-time communication.
b. High security with respect to reaching the called party *within a certain predetermined time* (but not immediately).
c. High security, with respect to reaching exactly the *right destination*. We cannot tolerate mistakes in call handling on the network.
d. High security regarding bit error rate (*faultfree connection*).
e. High security with respect to minimizing the risk of *wire tapping*.
f. High security with respect to having *100% availability of the network*.
g. High security with respect to having *100% availability of the called party*. (That is, availability during call setup and no risk of losing contact once the connection is established, for example, during a file transfer.)
h. *Handling mass-calling situations*.

These items place great demands on networks; however, not all of them must be fulfilled for every connection. Let us examine each scenario and identify the demands for each case.

1. *Fire department call*. For the fire department call, demand a (real-time communication), demand c (reaching the right destination), demand f (availability of the network), and demand g (availability of the called party) are the most urgent.

   It is important to reach the destination without delay. This includes, of course, 100% availability of both the network and the called party. Because voice is the method of communication, demand d is not important; that is, we can tolerate a certain amount of bit error. We also do not need to worry about things like demand e (wire tapping), and so on.

2. *Money transfer*. When we transfer a large amount of money via the network, we cannot tolerate reaching anything but the correct address (demand c). This scenario also has other requirements that were not important in scenario 1. Demand d, a fault-free communication, is very important. Avoiding wire tapping (demand e) is another. On the other hand, we do not need real-time communication (demand a), as we can afford to wait seconds or even minutes.

3. *Pizza order*. Here, the only demand is for real-time communication (a), as this is a voice connection. Moreover, if communication fails, we can always wait a minute or call again later. We may even call another pizzeria, if we cannot reach the first one. In other words, it will not be a catastrophe if none of the demands are fulfilled. It may, however, be irritating and not very good for the image or business of the pizzeria or the network operator.

4. *Tender facsimile*. If we are faxing a tender only minutes before the deadline expires, it will be a disaster if we do not reach the right destination (demand c). We almost require real-time communication; a delay of 10–20 seconds can be tolerated, but not more (demand b). Demands f and g (availability of both

network and called party) are important in this case. Demand e (avoiding wire tapping) could also be of very great interest.

5. *Computer mail.* There is no demand for real-time communication (a) here. Fulfilling demand b (reaching called party within a certain time), demand c (reaching the right destination), and demand e (avoiding wire tapping) is enough in most cases.

6. *Televoting session.* In this case, we must be able to carry out mass-calling (demand h), which is not always easy to do without disturbing other calls. When overload situations occur, it is important to:

- Ensure, as far as possible, that high-priority traffic originated at the same time is not affected. *In an ideal network, high-priority traffic should not notice the overload.*

- Minimize the damage for traffic with more moderate demands, that is, ensure that *as much as possible of the normal nonprioritized traffic still continues as normal.*

- *Minimize the duration of the overload,* including recovery time.

7.3 NETWORK TRAFFIC SOLUTIONS

If we translate the demands described in Section 7.2 into functions that can fulfill them, we can identify network needs for

- A priority function that allows certain calls to pass before others. Priority is essential for letting urgent traffic pass before other traffic. Urgent traffic is generated most often in situations like scenario 1 (call for fire department), but also in some real-time traffic situations (demand a). Priority is especially important during high traffic load periods, for example, during the televoting session initiated in scenario 6.

- Redundancy, that is, at least two different alternative paths on the network or mated pairs of equipment. Redundancy is important for fulfilling demands b (reaching called party within a certain time) and demands f and g (availability of both network and called party).

- Handshaking functions between the calling and called parties, especially when several people have access to the terminals on both sides, and the capability to receive acknowledgment when sending a message. Acknowledgments are especially important for fulfilling demand c (reaching the right destination).

- A flexible and adaptive routing system, which is especially important for demand a (real-time communication), demand b (reaching called party within a certain time), and demands f and g (availability of both network and called party).

- Encryption capabilities. The need for subscribers to encrypt speech, facsimile, mail, and so on, will probably grow during the 90s. This is important for demand e (to avoid wire tapping). Encryption can be done on the network or at the terminals.

- Fault-free connections. Some traffic requires this, for example, data traffic. This is important for demand d (low bit error rate), of course.

- Overload functions (call gapping and windowing). When a mass-calling situation (demand h) occurs, we should be able to calm down the traffic in some way to avoid overloading parts of the network.

The following sections discuss these traffic solutions one by one, both in general terms and with respect to the demands raised in Section 7.2.

7.3.1 Priority

The general function of a priority system is to allow some traffic to pass before other traffic that appears at the same time. There are, however, some fundamental matters to consider when building a powerful priority system:

- Of course, the relation between prioritized and nonprioritized traffic must be relatively low to make the priority powerful. (If 100% of the traffic has high priority, there is, in fact, no priority at all.)
- Another obvious matter that is often overlooked is the fact that the priority system must be implemented throughout the entire communication system. Otherwise, it may be of no use at all. For example, it is of no use to give a certain call priority on the local exchange or on the transmission system, if it is not also given priority on the transit exchange, on the local exchange of the called party, and so on. Considering the example of the call to the fire department, it is easy to see that unavailability of the local exchange of the called party during call setup could make all other priority systems useless.
- Sometimes more than one priority level is necessary; for example, a four-level system is often very powerful. With four levels, we can reserve the highest level for traffic of extreme urgency, like calls to the fire department. The lowest level could be used for traffic having almost no time demands at all. For instance, a network operator's maintenance traffic, counters for traffic management and statistics, and so on, could be sent at night or during other low-traffic periods. That leaves two priority levels for traffic with normal time-related demands.

Extra benefits could be gained if capability was provided for

- Raising the priority level of a message, if, for example, it waits longer than a specified time for handling.
- Reserving certain resources only for use by high-priority traffic. This could be done in more than one way. For example, certain resources could carry the stipulation that they are only for use by high-priority calls, or lower-prioritized traffic could be rejected if the load reached, for example, 80% of total capacity.

Any priority that exists today is usually directly associated with a certain network access point (NAP). Remember the distinctions in Section 6.2.2 concerning the calling party: the person's identity, the person's function, the terminal identity, the terminal function, and the NAP from which the call came? The priority could in future differ according to the person's role, the terminal's role, and the location of call origination (the NAP). Of course, we should also be able to assign different priority levels according to combinations of the different roles and the NAP. In Section 6.2.2, we looked at the example of Mr. Smith, who has several numbers, a personal number and two numbers for personal functions—one as general manager of the ABC company and one as the chairman of the local golf club. In future, he might be assigned different priorities when using the network based on when, from which NAP, from which terminal, and in which role he uses the network.

Obviously, a high priority will cost more. For example, if Mr. Smith uses the network from his office in his role as general manager, he is assigned a high priority, but if he uses the network as a private person, he is assigned a priority just like everyone else. Or perhaps the priority does not depend on his physical location, but instead on the time of day. In other words, he receives high priority only during working hours and normal priority at other times, independent on his physical location.

It should also be possible to alter priority between predetermined levels by using the customer control interface in combination with a password. Another benefit for customers is the capability to choose to use priority or not on a call-by-call basis.

7.3.2 Redundancy on the Network

In cases of high traffic on certain routes on the network as well as of failures on exchanges, service switching points (SSPs), service control points (SCPs), signaling transfer points (STPs), and so on, it should be possible to have at least one alternative way to communicate. In other words, the ideal situation is to have at least two ways to route a call at every point in communication. Of course, for economic reasons, duplication at every point is not possible. However, for vital functions that are common for many subscribers or services, a duplication or a backup is necessary. Redundancy could also be an easier way for network operators to provide high priority (described previously), as they can duplicate resources that normally are under heavy pressure from high-priority traffic.

In the case of intelligent networks, a very useful way of creating redundancy is to duplicate the SCPs by allowing two SCPs, preferably at different physical locations, to act at every moment as exact copies of each another. This is called *mated pairs of SCPs*. The two SCPs in a mated pair must *always* contain *identical data* and must be *updated simultaneously* when a service is introduced or changed or when data for a subscriber or customer is changed. This includes changes made by the customers themselves through the customer control interface. Mated pairs are covered in Section 7.4.2, which deals with future network structures.

7.3.3 Handshaking Functions and Acknowledgments

Before transferring the money, as in the transaction described in scenario 2, we must be 100% sure that we have reached the correct destination, which is why a handshaking function or a password sent back on the network would be of great value. After we have sent the money, we also want a confirmation of receipt from the other party via the network. The same is true for sending other important messages, such as facsimile, computer-to-computer mail, and images, through the network.

In Section 6.2.2, we discussed the possibility of having the calling party's identity transferred to the called side and the called party's identity transferred to the calling side. If it is the personal identity or the personal function that is transferred, both the calling and the called party will know who they are communicating with. This identity transfer could be used as a handshaking function.

7.3.4 A Flexible and Adaptive Routing System

The routing function for an IN-based service is only partly carried out by the SCP today. The SCP determines the destination for that setup, that is, where the called subscriber is located, but does not determine the path in the network. The path is determined, at least in the early stages of intelligent networks, in the same manner as any call setup in the network, that is, by the normal routing functions.

I believe that in future, the intelligent network will have the power and the capability to take over all routing in the networks, including determination of the path through the network to the destination. Put simply, I believe that the destination determination carried out by the intelligent network will merge with traditional routing to become a single system for traffic control and routing, handled by the intelligent network. How this can be carried out will be discussed in Section 7.5.3.

7.3.5 Encryption

To reiterate, use of the network has changed drastically. From having once provided voice communications, primarily for households, it has become a vital instrument for business. Many businesses are becoming more and more dependent on the availability of secure, fast, and reliable communication via the telecommunication network. As the number of business transactions continues to grow, so does the demand on networks for privacy and security. One important facility we should mention in this context is the encryption facility. As the name implies, this facility transforms a message, which can include speech, facsimile, mail, and so on, so that it cannot be read by anyone except the receiver (the terminating local exchange or the called party).

On a network, we encrypt messages primarily to avoid the risk of wire tapping, which is when someone *intentionally* tries to get unauthorized access to information during a transfer. Encrypting a message ensures that it can only be read by the receiving party. However, just encrypting messages does not provide users with 100% security. Other risks besides wire tapping must also be considered. For example, it is quite common for unauthorized personnel to gain access to information through *carelessness* at the sending or the receiving location. Problems can occur even when encryption is used resulting from users neglecting to encrypt *all* messages or from carelessness with encryption keys, passwords, facsimiles left in public places, and so on. Of course, when important documents are left where anyone can find them, the encryption facility is of no use whatsoever.

Encrypting facilities can be offered by the network provider or included in the customer premises equipment (CPE) at the customer's location.

7.3.6 Fault-Free Connections

In speech communication, bit errors causing noise on the line do not normally pose a problem. Usually, we do not even notice them. However, when sending a data file via the network, we must have a higher degree of fault-free connection, as even a low bit error rate could cause problems. We must remember, however, that there are several ways to correct faulty messages. One way is to have a good protocol, whereby faults in bits are corrected at the receiving party as soon as they appear, or the receiving party asks the sending party to retransmit if a fault occurs. Another way is to choose only high-quality connections on networks and avoid connections with bad transmission. In fact, all networks contain good and less good (new and old) connections. The good ones are often built with modern technology, such as optical fiber and stored program control (SPC) exchanges. The other ones are often old and use analog equipment—analog exchanges with transmission via copper cables.

Consequently, in the same way that a good connection places less demand on the protocol, a good protocol can compensate for a bad connection. But using a good protocol—assuming that we can—for all calls places a great demand on CPU performance in the exchanges, which is why this is not always a good choice. (We may get overloads, long delays in call setups, and so on.) However, in cases—a particular call or for a particular customer—in which we have an extremely high demand for fault-free connections, we should use a routing system that only selects high-quality connections that are controlled by a good protocol.

7.3.7 Overload Functions (Call Gapping, Windowing)

The call gapping function only allows a predetermined number of calls to pass per second between two points. This predetermined number may never be exceeded. If the demand for traffic is higher than that number, only the predetermined number are sent per second. Remaining calls are temporarily stored. One common use of call gapping is for communication, via the signaling network, between the SSP and the SCP on an intelligent network, for example, for the purpose of translation of a premium rate or televoting number on the SCP to its destination number. The SCP, in particular, risks an overload in the case of mass calling, as many SSPs simultaneously attempt to reach the SCP. If call gapping is introduced on all connections from the SSPs to the SCP, we can control the traffic flow from each SSP, but we can never be sure that the SCP is not overloaded.

The windowing technique is a more sophisticated method of controlling traffic flow. While call gapping is a static method, that is, only a predetermined number of calls are ever allowed to pass, the windowing technique takes into account the traffic load on the

SCP and so works dynamically. When calls are being sent from SSPs to the SCP, at the same time, we are waiting for return calls from the SCP to the SSPs. Windowing determines how often calls can be sent to the SCP based on the number of outstanding calls, that is, the number not answered yet. If too many outstanding calls accrue, sending is temporarily stopped, until outstanding calls are answered. When a certain number of answers have been returned, sending is continued. When windowing is used from all SSPs to the SCP, the actual load on the SCP decides the traffic flow from each SSP, thus risk of overloading the SCP is less than when call gapping is used.

7.4 IMPACT ON FUTURE NETWORK STRUCTURE

Almost all traffic carried on the telephone networks during the last decades has been real-time voice communication, that is, normal telephone calls between two people. The greatest demand from users, especially the calling party, has been to obtain a quick response from the network after dialing a number (a few seconds) to find out if there is a possibility to reach the called party or not. If congestion occurs on the network or if the called party is busy, we want to know that immediately after dialing the last digit. Even a ten-second wait is intolerable. Once a call has been established, however, we do not notice or care about single bit errors, as when we are speaking we can understand each other anyway.

If we look ahead 5–10 years, many experts forecast a major change in the user pattern of subscribers in PSTN. Traditional real-time voice communication will still be used a lot, of course, but it will probably only account for about 50% of all traffic. The rest will come primarily from facsimile and mail (computer and voice) traffic. The narrowband integrated services digital network (ISDN) will also have grown, but to far from the same size as the PSTN, of course. Broadband communication, which is primarily used for business, education, and leisure, will also grow.

Many experts believe that ISDN access will replace a great deal of existing PSTN accesses by this time, and that ISDN will be heavily used in the United States and Europe. Personally, I do not believe this will be the case. There are some very concrete facts that support my view:

- ISDN access forces an extension of a new physical access (termination) to every new customer. This is a very expensive task that I do not think is possible, at least not during the economic recession we are experiencing in the beginning of the 90s.
- The possibilities offered by facsimile and computer mail communication on the PSTN provide many of the benefits that were supposed to be offered by ISDN.

However, I do believe that broadband ISDN communication will expand, as there are only a few alternatives offered in PSTN. When there are alternatives in PSTN, you often have to multiplex numerous 64 Kbps channels to a higher bandwidth, which is a rather complicated and expensive alternative.

If what we discussed above is true, in 5–10 years we will end up with a PSTN that carries two major types of traffic, with about a 50–50 division between them:

- Conventional voice communication, like we have today, with a demand for a 1–2 second connection to the called party.
- Facsimile and computer or voice mail traffic, with demands for response times of 5–10 seconds or more and for such features as multicast capabilities and called party receipts.

ISDN accesses, both narrowband and broadband, will increase in parallel, however, and the ability to cooperate with the major PSTN network will be mandatory for the minor ISDN. I believe that ISDN cannot survive without that cooperation from PSTN.

The SCPs, first built and used in the PSTN, may in the early phases of ISDN also be used to handle traffic (especially number translation services like freephone and premium rate) between

- PSTN and narrowband ISDN,
- PSTN and broadband ISDN,
- The ISDN networks themselves,
- Narrowband and broadband ISDN.

Of course, communication capabilities are determined by the weakest party, and this determination could be managed by the SCPs on intelligent networks.

7.4.1 New Network Topologies

The strict hierarchical switching networks we have had traditionally often have from four to six levels of exchanges. These can range from a very small analog local exchange at the bottom to the digital transit and/or international exchange at the top. Introduction of SPC exchanges and digital transmission, however, has led to

- Increasingly larger exchanges than before,
- A reduced number of switching levels in many networks to two or three.

Consequently, when an intelligent network is introduced, we often have two or three levels on the switching network complemented by some SCPs that control IN-based services. In this first phase, the SCPs are not working hierarchically, which means they only form one level of intelligence or one level of service logic on the network. The trend in the switching networks is to further reduce the number of levels to one or two, usually by building combined exchanges for many purposes.

Where intelligent networks are concerned, we can spot another trend. The number of SCPs is starting to grow as the number of IN-based services increases. Not only is the number of IN-based services growing, the number of service providers and service subscribers that use a single service is also growing. This growth results in an increase both in the signaling traffic going to the SCPs and in the load on the SCPs themselves. In the area of service, we will see a distinction between services with relatively moderate demands on intelligence, such as number translating services, and services demanding more advanced intelligent network functions. The latter may include sophisticated functions for customer control, advanced charging, and so on. Demands are also being raised for an IN-based service to reach across more than one network, which is why matters like cooperation between SCPs are becoming increasingly interesting.

This evolution will most likely result in demands for more than one level of SCPs in the network. Consequently, I believe we will see a further decrease in the number of levels on the switching networks and an increase in the number of levels on the intelligent network. (See Figure 7.2.)

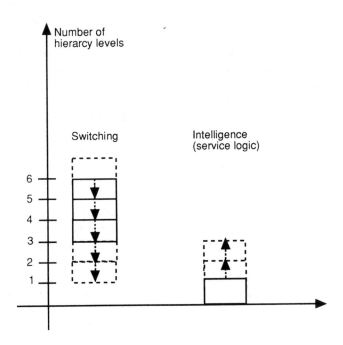

Figure 7.2 Hierarchy evolution in both switching and service logic (intelligence) networks.

I also believe that the signaling protocol we are implementing in the early 90s—basically, for communication between SCPs and SSPs—will be divided in future into two or three different protocols. As we have just seen, the demands of the services in future will have a great span in terms of intelligence and security requirements. At one end of the spectrum, we will only need a number translating capability for televoting, while at the other end, we might need advanced call setup functionality, perhaps based on artificial intelligence (AI). It would be a great waste of resources to design all future protocols for the worst-case scenario. This is why I believe there will be two or three different protocol families with different functionality and different security levels, each suited for different purposes.

We examine possible cooperation between intelligent network platforms and hierarchies of SCPs in Section 7.6.

7.4.2 Mated Pairs of Service Control Points

In Section 7.3, we saw that a main demand of future networks regarding the concentration of intelligence logic for a service in the service control function (SCF) is to have at least two identical SCPs, physically separated for security purposes. See Figure 7.3. The two SCPs would work as a mated pair, that is, they would be exactly identical at all times, like SCPa and SCPb in Figure 7.3. Both SCPs would always contain the same subscriber and service provider data and one is appointed for the normal handling of the service (the executive). The second SCP (standby) is called only if the executive cannot be reached. When a change is made, including changes made via the customer control interface, the service management point (SMP) would have to update the two SCPs simultaneously. In Figure 7.3, both SCPs have a copy of the two services Z and W for service provider X. However, SCPa is the executive for service Z and the standby for service W. The opposite is true for SCPb.

If service provider X wants to use the customer control function to make a change in service Z, for example, to change the routing or the passwords for a customer, this could be done in two different ways:

1. The service provider accesses the SMP to make the changes. After that, the SMP updates both SCPa and SCPb in the mated pair. The provider gains access via an intelligent terminal or maybe, in some cases, via a dual tone multiple frequencies (DTMF) interface.
2. The service provider accesses SCP-a, which is the executive SCP for service Z, and makes the changes. (This procedure is for more simple tasks, like changing PIN codes, and so on.) SCP-a then updates the SMP, which updates SCP-b. This type of access is mainly done via a DTMF interface.

The SCPs must be updated simultaneously or, in the second case, without any substantial delay in the updating of SCP-b (the standby). The standby SCP must always be ready to take over all traffic for service Z for service provider X if the executive SCPa fails for any reason.

For traffic handling, mated pairs of SCPs could offer different solutions:

- One SCP is the executive and handles all traffic, while the other remains passive and on standby. If the executive SCP fails, the standby takes over the traffic immediately (redundancy).
- Both SCPs are executive, either for different types of traffic (different services) or for the same services in different geographical areas (loadsharing).

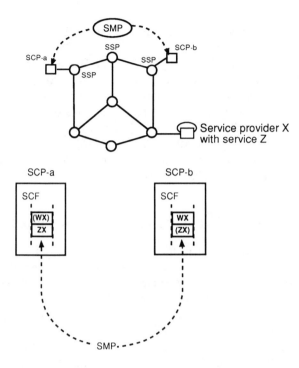

ZX = Service logic for service Z,
owned by service provider X, is duplicated.

Figure 7.3 Mated pairs of SCPs.

- Both SCPs are executive, either for different types of traffic (different services), as in Figure 7.3, or for the same services in different geographical areas, and they are both also on standby for each other. Should one fail, the other will take over its traffic (redundancy and loadsharing).

The last solution is used quite often. In the example in Figure 7.3, SCPa is the executive SCP for service Z and SCPb is the executive for service W. In the same way, SCPa is a standby for service W and SCPb is a standby for service Z. During normal traffic conditions, we have loadsharing between the two SCPs, as they each handle one service. However, if one fails, the other one will be able to take over all traffic.

One problem we need to address when discussing mated pairs is how the handover from one SCP to another should be carried out in the event of failure. A very handy way, which was used during the first years of intelligent networks in Sweden, is simply to include the SCPs in the normal routing scheme on the transit network. With this method, all transit exchanges—which also served as SSPs in their routing scheme for service Z—have SCPa as the first choice and SCPb as the second. Now if a call is made for a number translating service (service Z), for example, from an SSP to an SCP, the SSP would attempt to reach SCPa in the same way that a normal call setup is made. Normally, SCPa would answer and service Z would be carried out. However, if SCPa did not answer the SSP for some reason, which could be due to an overload, SCP failure, or some other type of network failure, the SSP would automatically reroute to the second choice in its routing table (SCPb), and the service could be carried out. This solution is based on the national telephone user part (TUP) in CCSS No. 7. Two new signals were simply added to the TUP for handling the communication between SSP and SCP.

It will be possible in future, however, to achieve handover by using the SCCP global address facilities in CCSS No. 7.

7.5 ALLOCATION AND OPTIMIZATION OF RESOURCES ON NETWORKS

If we examine the intelligent network from a longer perspective, it becomes obvious that the technology can be utilized as a network with a large amount of intelligence—*for both service control and call control.* In some cases, this intelligence will be centralized in a few points, and in others, it will be distributed throughout the network. Besides SCPs, intelligent peripherals (IPs) will be extensively used, both as standalone controls for single services and as complements to the SCPs. At the customer's own location, customer premises equipment (CPE) will handle very unique services or share the handling of services with SCPs and IPs. We will find that an increasing number of service data points (SDPs) are supporting the SCP. Finally, the IN-based logic of telecommunications networks will interwork with outside databases owned by large companies, such as banks and finance companies.

Because of these increasing demands, networks operator must continuously keep an eye on the *network economy*, which means providing customers with exactly what they are paying for at a minimum cost. The cost depends heavily on such matters as the provisioning of resources on the network. Resources must be optimized in the right quantity and allocated to the right places. They can also be allocated to different places in different timeframes.

7.5.1 Resource Allocation and Optimization for Announcement Machines

For some resources, optimization and allocation were important issues for years prior to the introduction of intelligent networks. One example of such a resource is the announcement machine used by network operators.

Announcement machines are used for a variety of purposes, the most common of which are messages to subscribers that announce a network failure or congestion or are for interception purposes. Announcement machines can be implemented on a low level, such as at every local exchange, or higher, at an exchange on the transit level. Usually, the expected traffic load in combination with the network economy determine the implementation chosen. As announcement machines are often (or at least used to be) very expensive, the network operator seeks to minimize their number. On the other hand, using only a few announcement machines results in very long routes to send a message to a subscriber. Consequently, it is an important economic factor to optimize (minimize) both the number of announcement machines and the length of the routes on the network.

An example illustrates how the employment of announcement machines can be optimized. Some IN-based services need to provide a message to customers, for example, during the setup. (Of course, non-IN-based services also send announcements, and the same announcement machines can be used by both services, but, for the purposes of our example, we will consider only IN-based services.) Chapter 3 describes intelligent network solutions that depend on whether a signaling and speech connection is set up to the SCF or only to a signaling connection. The first solution featured a service switching and control point (SSCP) and the second employed an SCP. Figures 7.4 and 7.5 (which are modified versions of Figures 3.3 and 3.5) show examples of different announcement machine allocation methods.

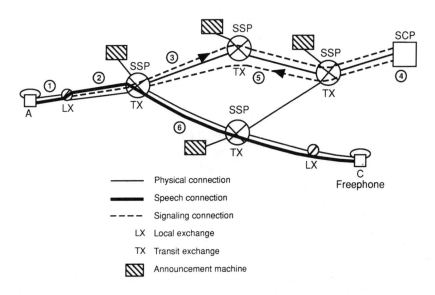

Figure 7.4 Announcement machine allocation with the SCP solution.

In Figure 7.4, there is only a signaling connection, no speech connection, set up to the SCP, which is where the SCF is located. In Figure 7.5, there is an SSCP, which therefore forces the call to be routed, both speech and signaling, through this node. Now, what if we need to make an announcement during the setup of an IN-based service? This announcement is preferably sent by the first SSP reached in the setup in Figure 7.4. In Figure 7.5, the best way is for the announcement to come from the SSCP. In both cases the message is controlled by the SCF, which is located on the SCP in Figure 7.4 and on the SSCP in Figure 7.5. Consequently, in the SCP solution, we need to install one announcement machine at every SSP. In the SSCP solution, we need to install only one announcement machine at the SSCP itself. This assumes that capacity is sufficient, of course.

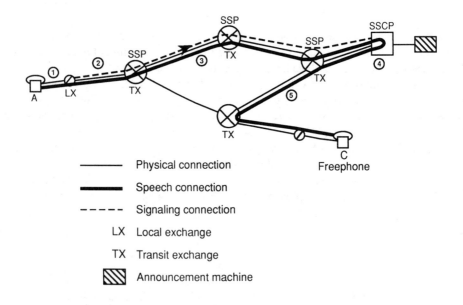

Figure 7.5 Announcement machine allocation with the SSCP solution.

But, as we saw in Chapter 3, the SSCP requires all the speech channels to pass the SSCP, which is poor network economy because it requires longer routes, in total, for every IN-based service compared with the SCP solution. To summarize, in the SCP solution we save speech channels on the network, but the cost is higher for announcement machines. In the SSCP case, it is exactly the opposite.

In the future there will probably be a move towards SCPs rather than SSCPs, as the use of signaling according to CCSS No. 7 with the INAP protocol expands. This will probably result in announcement machines being distributed on the network, yet controlled from SCPs via the signaling network. Also, announcement machines distributed on the network, built up on phrases (words), may be remotely controlled by SCPs. The SCP would determine the order of the words in each message; this determination would be sent to the SSP; the SSP would send the real messages to customers.

There are other interesting matters concerning the allocation of announcement machines. Maintenance considerations can make one distribution method more viable than another, for example, if we want maintenance staff to be able to reach the machines within a certain period of time. On the other hand, today and especially in future more and more maintenance will be carried out remotely.

7.5.2 Resource Allocation and Optimization for Service Logic

Service logic allocation is discussed in Chapter 2, Section 2.2.3, which describes the future evolution of the intelligent network. Here, we continue the discussion of service logic allocation, using two examples.

In the first example, a customer of a future personal communication network (PCN), with full mobility, has a personal number, which is probably combined with a personal service profile. The personal service profile includes the IN-based services the customer has subscribed to, many of which are customized to meet her demands. Moreover, she has personal data stored with her services, including personal abbreviated numbers, accounting data, and so on. All of this data is stored on an SCP, which we will call the *host SCP* for the customer.

If the customer travels around in the network and ends up at a location outside the area covered by her *host SCP*, she still wants to be able to use her personal service profile. One solution to this problem is to set up a signaling connection to the host SCP to check the service profile for every call made. However, this not only slows down communication (call setup) for the customer, it is also a waste of signaling resources for the network operator.

A far better solution would be if, when the customer came to a new place and was identified by another SCP, the latter would simply retrieve a copy of her service profile from the host SCP. In that case, the service profile would have to consist not only of the services the customer can use, but also the changes she is allowed to make via the customer control interface.

In this example, use of signaling connections is minimized. The host SCP keeps a master copy of the profile, and any changes made by the customer via the customer control interface while she is outside of the host's area are reported back to the host SCP. The changes made must, of course, stay within the limit of what the customer is permitted to do. If in doubt, the SCP in the visited area must ask the host SCP for advice. When the customer moves on to a third location, the host SCP is responsible for removing copies of the service profile on the SCPs in old areas (for example, at the moment the PCN customer registers herself at the third location).

A second future-oriented example involves a customer who may not have all services available directly, because there may now be several hundred or several thousand on the network. A single customer simply does not need all of them. But if he suddenly wants to use a service that is not available in the place it is needed—on the local exchange, on the SCP, or even at a terminal—he should be able to demand it. When it is demanded, it is retrieved from a "database" of services, that is, downloaded to the local exchange, to the SCP or to the terminal, depending on the implementation method for that specific service logic.

Figure 7.6 (which is identical to Figure 2.7) illustrates how service logic in future could be given *on demand*. The figure shows the downloading of service logic (intelligence) to a local node.

7.5.3 The Evolution of Resource Allocation and Optimization

Sections 7.5.1 and 7.5.2 describe two examples of needs for resource allocation and optimization. This section examines the evolutionary phases resource allocation and

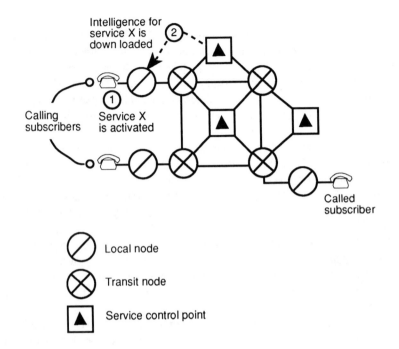

Calling subscribers

Service X is activated

Called subscriber

⊘ Local node

⊗ Transit node

▲ Service control point

Figure 7.6 Downloading service logic to a local node

optimization went through prior to the introduction of intelligent networks and looks at some possible future evolutionary trends:

Phase 0. If we look at the evolution of SPC exchanges, we know that they consist of software and hardware. The hardware is the physical equipment inside the node, and it includes signaling resources to the next exchange on the network, resources for connection of subscribers, announcement machines, and so on. The software is the *logic for controlling the exchange* (the node), that is, the hardware.

The software and the hardware are combined to form a *system* that is the node itself. The node handles both normal calls and the services allocated to it. The system has been optimized in such a way that both the hardware and the software have been developed with a special emphasis on economy. We have carefully calculated the amount of physical equipment needed for signaling to minimize cost. The software has also been optimized through an attempt to find a good structure with good interfaces between modules, to minimize the CPU load.

We could also characterize phase 0 as concentrating on the *hard and soft resources* on the node. It is important to remember in the following discussion that the soft resources control the hard resources, or in other words, that together they make up a system.

Phase 1. If we look at the location of a node on a network, we see that routing schemes were introduced long ago to find the best or at least a satisfactory route through the network. The fundamental criteria was based on *minimizing* the use of (hard) equipment on the network, that is, to pass through as few exchanges as possible and to minimize the use of transmission equipment.

This is what is done, prior to the introduction of the intelligent network concept on a normal network. This is a way *to optimize the use of hard resources* on a network. Routing can then be regarded as a soft resource, while nodes, transmission, and signaling are the hard resources.

So, before the intelligent network concept, we have a situation with a system for *optimization of both hard and soft resources inside the node* (phase 0) and a system for *optimization of hard resources on the network* (phase 1). What is missing is a system that deals with optimizing soft resources on the network level.

Phase 2. Introducing the intelligent network concept for the purpose of controlling services, which, as we have seen, is the first step for IN, offers a partial solution by also providing a method of *optimizing soft resources for service control on the network level.* Introducing the intelligent network concept makes it possible for control logic to be moved to locations other than where the services themselves are performed. Usually, logic for a service is centralized on one SCP for all subscribers, which is an excellent way to optimize the allocation of that logic.

This is the status today in most networks that have implemented the intelligent network concept: Systems exist for optimizing and allocating hard and soft resources for nodes and for hard and soft resources for networks with one exception—the soft resources for call control. What are the next steps of evolution? As we move forward, we find different evolution paths that do not follow each other strictly. They can be introduced in parallel, in sequence, or as individual alternatives.

Phase 2a. One possible step for the future is to extend the routing capability (soft resources) provided by the intelligent network to *include normal basic telephony calls.* Or, in other words, to let the SCPs on the intelligent network control all routing on the overall network too. While this could be seen, in the early phases of intelligent network implementation, as too large a step, we should consider the following scenario.

By the end of the 90s, perhaps 50% or more of the switching activity on the network will emanate from services, that is, from switching on the transmission network and on the signaling network, with the latter mostly for messages between SSPs and SCPs. The other 50% will emanate from normal basic telephony.

Of the 50% of switching emanating from services, the majority will come from IN-based services controlled by SCFs allocated on SCPs. These services will have features, such as time- and origin-dependent routing, advanced customer control (Section 6.3.2), or numerous advanced features we cannot envision today.

Subscribers who use the advanced features of the IN-based services will certainly need to use them in combination with normal calls. For example, if I can use time- and origin-dependent routing with advanced customer control for the freephone number going to my business location, I will probably want to use the feature with my ordinary number as well, that is, for the basic calls I make. Furthermore, I will want to use them in the same way.

For network operators, there are two driving forces behind letting the intelligent network handle the normal routing of basic calls in addition to handling routing of services:

1. Demand from customers and subscribers,
2. Network economy.

If nearly 50% of routing is controlled in one way and rest is controlled in another, it is vital to harmonize the two, that is, to control 100% in the same manner. To choose the intelligent network becomes very easy, as the intelligent network solution will offer many advantages over conventional routing.

Phase 2b. Most networks set up the path through the network for the routing scheme in a *static* manner, meaning the path between exchange A and B is always chosen (allocated) in the same way, regardless of the traffic situation. If the choice of a route could also be dependent on the actual load situation in the network, we could create a *dynamic* way of routing. The SCPs in the intelligent network offer an overall view of the traffic load on the network. They alone have knowledge of the actual traffic from the IN-based services, and, as we saw in phase 2a, will perhaps know about traffic loads for basic calls in future as well.

Consequently, by allowing all SCP information about actual traffic loads (to and from which nodes traffic flows) to be reported to the SMP in future, the latter obtains an overall view of traffic on the network. The SMP can then calculate the momentary load on each exchange and each route between exchanges and report this to all SCPs. The SCPs, in turn, can use this information for routing purposes, for example, to avoid heavily loaded exchanges and routes.

The reports to and from the SCPs and the SMP form a system of *dynamic even-load distribution* over all the equipment (hard resources) on the network. Such mechanisms are already developed in the traffic management network (TMN) area. This suggests that we need *either* closer cooperation between the intelligent network and the TMN in future *or* clearer requirements from the intelligent network for the TMN to be able to introduce dynamic routing.

Phase 3. In this phase, we consider interworking between networks. In the future, when services become global, intelligent network systems must be able to interwork in the same way that networks interwork for normal calls today. However, if we look at service interworking, we can identify two types of services: services that are controlled by the first network's own SCPs and services that function across networks and need for some sort of *global common control*.

Evolution is undoubtedly headed towards a concept whereby a common SCP handles services extending across more than one network. This SCP, which we will refer to as the *common service control point* (CSCP), will handle those global services. By global, we mean services that function across two or more networks. Aspects of CSCPs are covered in the next section.

7.6 INTERWORKING BETWEEN INTELLIGENT NETWORK PLATFORMS

As Section 7.5.3 mentioned, interworking between intelligent network platforms will be necessary in future, for example, when a service must cross more than one network. I do not intend to describe such a technical solution fully, but will instead approach it from a macro level. There are three basic scenarios for interworking between intelligent network platforms. Figure 7.7 shows three different networks—each with its own intelligent network (that is, with SCPs and an SMP)—and how they could cooperate in future in delivering common services.

In 7.7(a), there is no technical cooperation at all, only administrative cooperation. Service creation could be concurrent across networks and implementation could be carried out at the same time in each network's SCPs via the SMPs. (The SCPs and SMPs in networks are used for both internal network services and common services.) Timing is of the essence here, that is, services must be opened simultaneously in all networks.

In 7.7(b), the SCPs in each network are still used for both internal and common services. The main difference, however, is that the common service management point (CSMP) manages common services, while internal SMPs manage internal services, as in 7.7(a). One of the problems with this solution is that there will be two SMPs working with the same SCP. But the most probable solution is that the CSMP will communicate to the SCP via the SMP, the latter having total control over all communication with the SCP. The SMP can therefore avoid, for example, sending contradictory messages to the SCP.

The scenario in 7.7(c) separates the common systems for both the SCP and the SMP. We now have four systems: one system for each of the three network operators for internal services and a fourth system with a CSCP and a CSMP for the common services. A subscriber may use different prefixes to distinguish between internal and common services, of course, thus allowing the SSP to decide whether to interact with its own SCP or with the CSCP. But if we want to have the same prefix for a service regardless of which SCP handles the service, we will need to route all the services to the SCP in our own network. This SCP must then distinguish between its own services and services that are to be carried out by the CSCP.

Which of these solutions that will be applied in future is very difficult to predict. Network operators will probably negotiate the decision on a case-by-case basis, depending on the condition of individual networks involved.

a)

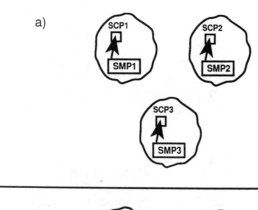

b)

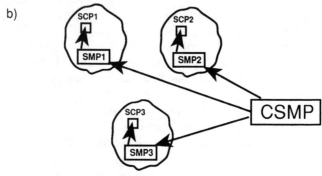

c)

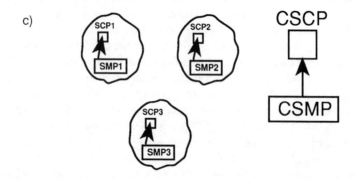

Figure 7.7 Possible global intelligent network–based service provisioning: a) no technical link manages common services; b) a common service management point (CSMP) manages common services; c) a common service management point (CSMP) and a common service control point (CSCP) manage common services.

Chapter 8

Building, Operating, and Using Large, Complicated Telecommunication Networks

In Chapter 1, we made one observation and asked two fundamental questions about the evolution and future expansion of telecommunication systems. The *observation* was that large systems, such as telecommunication networks, tend to evolve in waves. First, there is a period in which no visible results can be seen but a great deal of money is used to build a basic system. In the second period, which is not so costly from an investment point of view, there are visible results and return on investment. Later on, the cycle starts over with investment in a new basic system or in updating the existing system. (Please refer to Section 1.1.)

The *first question* raised in Chapter 1 singled out the *Large SYstem Dilemma* (LSYD) and the lack of solutions to this problem. The LSYD states that a large system will reach a point at which it can no longer expand. The key factor in this dilemma is the increasingly longer period of time required when demand rises to either implement a new system or upgrade the old system. (Please refer to Section 1.2.)

The *second question* raised in Chapter 1 was how to increase the use of services. We identified the need for user friendly interfaces and terminals, but found that this was not enough. Something was missing. The missing component was *user willingness* (UW). How can we change the situation from having network operators *offer* services to customer to having customers approach network operators *demanding* services? The customers, in this case, should include both service providers and service users/service subscribers. (Please refer to Section 1.3.)

In this chapter, we discuss the tendency in the evolution of large systems (observation) and both questions. We also consider the impact of the intelligent network concept and how it could help to solve these problems.

Evolution for the intelligent network will still take place in waves, but the waves will occur with a greater frequency than described in Section 1.1. The intelligent network will make it easier to find solutions to the LSYD and UW questions. LSYD is covered in Section 8.2 and UW is covered in Section 8.3. Section 8.4 is a conclusion.

159

8.1   LARGE INTELLIGENT NETWORK SYSTEM EVOLUTION

Figure 1.1 in Section 1.1 illustrates the impact of time on the up-to-date factor. When a large system is installed, ideally it fulfills the requirements placed on it. After a while, however, general demands increase because the technical evolution continues and other systems are constructed with greater capabilities. This means that the system becomes obsolete or the up-to-date factor of our system continuously decreases. Once it falls below a certain low level, the system becomes a liability rather than an asset to its owner. Ideally, we should upgrade or replace a system *before* it becomes a liability. However, if the system is very large and complex, these are very expensive and time-consuming procedures. In reality, most systems experience a period, before they are upgraded or replaced, when they are more of a liability than an asset. This discussion is applicable to building networks in general and to building services in the old (node-based) way in particular.

Figure 8.1 shows the difference between node-based service implementation and IN-based service implementation. For node-based implementation, more surrounding software is involved and thus a larger system is involved. For IN-based implementation, on the other hand, there are distinct interfaces to the platform and other services, as well as larger building blocks, which make it both faster and easier to update the service.

There is no direct dependency between the time required to introduce a service and the frequency with which new services can be introduced. However, if it takes substantially less time to introduce a service than before, as is the case with intelligent network technology, then the interval between new services is also almost certain to be shorter.

To summarize, the shorter the introduction time for services, the more often we can update them and the lower the risk that they will become old-fashioned and a liability (see Figure 8.1).

8.2   INTELLIGENT NETWORKS FACILITATE THE BUILDING OF LARGE SYSTEMS

In Section 8.1, we concluded that by using intelligent network technology to build services, we reduce the risk of a service becoming out of date and a liability to users. The short time required for introducing a service with intelligent network technology impacts on the cost as well. IN-based implementation is not as expensive as conventional methods because the time for design, testing, and implementation is shorter.

Let us look at how the intelligent network impacts on the solution of the LSYD, described in Section 1.2 with respect to both services and platforms. Section 1.2 identified three major factors that must be taken into account when an organization is considering improving an existing system:

1. The cost of improving the system,
2. The existing system itself,
3. The technological evolution in general.

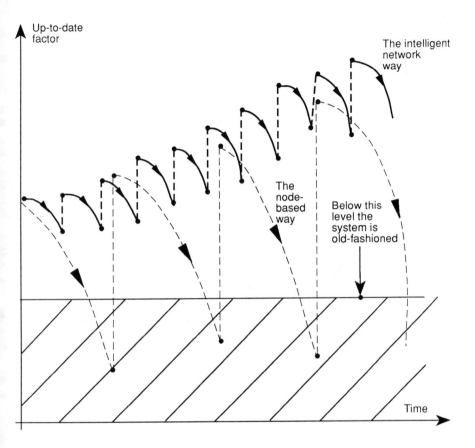

Figure 8.1 The up-to-date factor for intelligent network–based services.

In considering these factors, we can draw the following conclusions.

The Cost of Improving the System

For IN-based implementation, cost can be divided into two parts:

(a) Improving the intelligent network platform,
(b) Introducing new services on the intelligent network platform.

Where (a) is concerned, the conclusion reached in Section 1.2 is still true, that is, the larger and more complex the platform, the more costly it is to expand. Expansion of the platform includes the introduction of new signaling systems, a higher degree of redundancy, and an increase in CPU performance or in the security level, among other things. (This is shown in Figure 8.2, which is identical to Figure 1.2.)

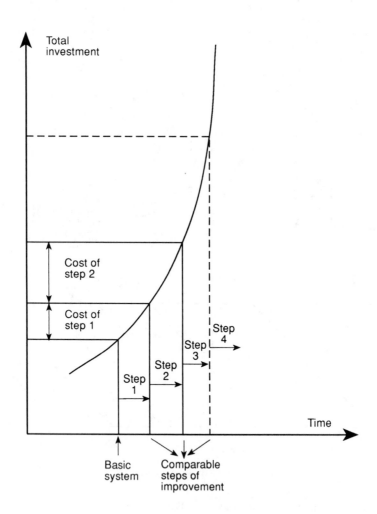

Figure 8.2 The cost of updating a system (a platform).

Where (b) is concerned, the cost for service introduction is drastically decreased when the intelligent network concept is used (Figure 8.4) compared to a node-based introduction (Figure 8.3). As Figure 8.4 also shows, the results are also positive for total investments in the long run.

The Existing System Itself

The same is true as for the cost of system improvement. Where the platform is concerned, we still face the problem identified in Chapter 1, that is, the fact that improving an existing

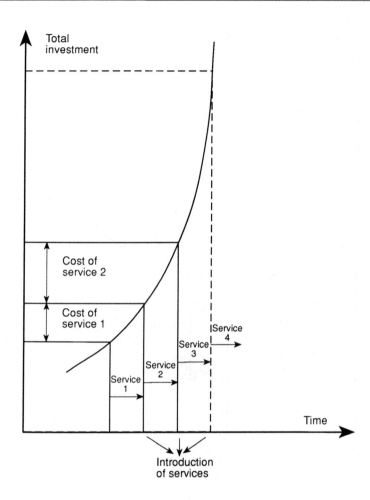

Figure 8.3 The cost of introducing a new service with conventional (node-based) methods.

ystem instead of building a new one is often a drawback. With regard to introducing new ervices on an existing platform, however, we find that doing this on an intelligent net-work platform eliminates the problem entirely or reduces it substantially. The reasons are hat services are built with large building blocks, that changes are needed only in a few oints, and so on.

The Technological Evolution in General

Figure 8.5 illustrates the life cycle of a node-based service at a certain point prior to the in-roduction of an intelligent network. (Figure 8.5 is identical to Figure 1.3 in Section 1.2, xcept that "service" replaces "system.")

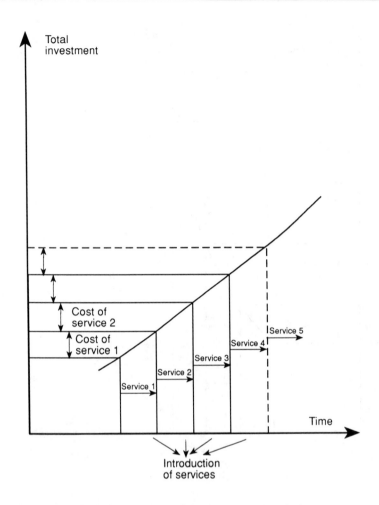

Figure 8.4 The cost of introducing a new service with intelligent network methods.

The life cycle is divided into three phases:

(a) The time required by a vendor to update an existing service or to develop a new one, or the time for a customer to order a service and have it introduced (Ta).
(b) The time during which the service fulfills user requirements (Tb).
(c) The time during which the service is still in operation but no longer fulfills user requirements, that is, the service is a liability (Tc).

Again, when discussing intelligent network platforms, we find the same trend described in Section 1.2. Introduction time (Ta) extends as evolution continues because systems (platforms) become greater and more complex every year. When discussing IN-based

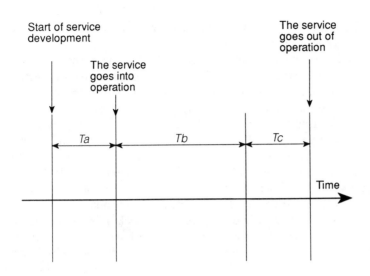

Figure 8.5 Life cycle of a node-based service.

ervices, we find not one but two interesting phenomena compared with node-based services. Please, see Figure 8.6.

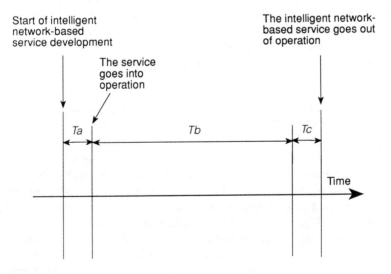

Figure 8.6 Life cycle of an IN-based service.

Ta is drastically shorter for an intelligent network introduction. In an intelligent network with the service controlled by a service control point, reached via a service switching point, *Ta* is still dependent on network size, but to a far less degree than before the intelligent network. The reason is as follows.

The updating of the SCP with new service logic is independent of network size, as the SCP will serve the entire network (unless capacity forces us to open a new SCP, that is improve the platform, which we can ignore in this case). Often the new service forces us to change access codes or routing patterns. Then, we must update all the SSPs, and this is network-size-dependent. In other words, the more SSPs there are to update, the longer it will take and the more it will influence *Ta*. Still, updating will go much faster compared with updating all switching systems on the network with service logic as is the case for node-based service introduction. Therefore, *Ta* (introduction time) is much shorter with the intelligent network than before, and it increases much more slowly, as the size of a network increases, than was the case before the intelligent network.

But, as has been stressed in other parts of the book, it is not only the technical aspects of a service that must be considered. So far what we have discussed is only true if *two conditions* are fulfilled, namely, *first,* that we find a fast and reliable instrument for testing service interworking, especially service interaction, prior to introduction on the network. The *second* condition is that the surrounding systems (management and administrative systems) can be updated in the same short time period as the technical system. Please refer to Section 6.4.

Tc probably decreases drastically with intelligent network introduction. How do we know this? *Tc* is the time during which a service is still in operation but no longer meets the requirements. Before the intelligent network, a service remained on the network for a long time because it took quite some time to create and introduce a new service. With the intelligent network, the time for finding and introducing a replacement for an old service is reduced. Consequently, the old service can be replaced more quickly by a new one and *Tc* decreases.

The situation will therefore be as shown in Figure 8.6 (compare with Figure 8.5). The main time in the life cycle of a service is now (as it should be) the time it is in operation (*Tb*) and fulfilling user requirements.

As we did in Section 1.2, let us add another dimension of complexity regarding both intelligent network platforms and IN-based services, that is, *Tg*, which is the time between the birth of two consecutive generations of the same intelligent network platform or of the same type of service. For intelligent network platforms, the same is true as was described in Section 1.2. See Figure 8.7, which is identical to Figure 1.4 in Section 1.2. But how does *Tg* (generation time gap) figure in when discussing an IN-based service, that is, how often is a new generation (an improvement) of a particular service demanded?

It is very difficult to say how often a new generation of a service is demanded. For some services, *Tg* could be the same as before the advent of intelligent networks, or even longer, but most probably it will be shorter. *Tg* will be determined primarily by market demands and by the particular service. *Tg* would likely remain equal or become longer when two or more services are competing (i.e., solving the same user needs). Here, it might be

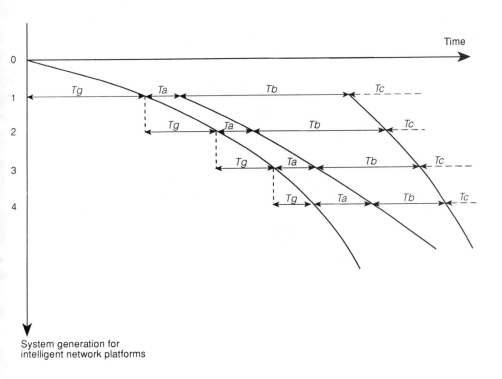

Figure 8.7 The evolution of intelligent network platform system generation.

that one of the services is replaced very often (low value of Tg) while the other or others stay for a long time on the network before being replaced. But, most often, and especially in the future, for the majority of market-driven services, Tg will be much shorter than before.

So, as Figure 8.8 illustrates, Ta does not tend to increase drastically for each new generation, as it did before intelligent networks. Moreover, Tb has been extended in relation to Ta and Tc, which have both decreased. (However, it is not certain that Tb has been extended in absolute numbers compared with before intelligent networks, since Tb is now largely dependent on market demands.)

Where service introduction is concerned, this answers the *first question* in the LSYD mentioned in Section 1.2. Regarding IN-based services, Ta does not expand too rapidly and the size of the system (the number of services) can be extended much more before Ta reaches the size of Tb. We have then solved one part of the LSYD problem. Thanks to the intelligent network, we can continue to increase the size of the system considerably before the LSYD appears.

The *second question* was the risk that Tg would become less than Ta, that is, that new generations of systems would appear faster than the time needed to develop and implement them. As we have seen, Ta has been drastically decreased when moving from old

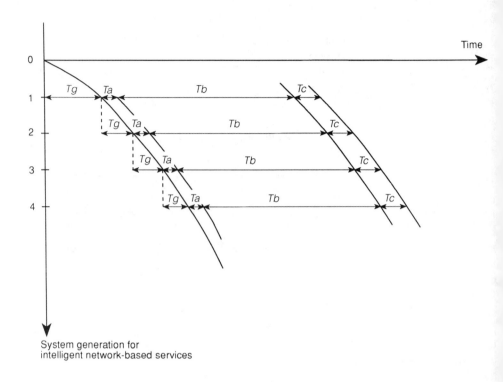

Figure 8.8 The evolution of intelligent network–based service generation.

basic implementations on nodes to an intelligent network, and Tg is dependent on market demands.

As in the first question, the drastic decrease in Ta substantially reduces the risk that Ta will exceed Tg for IN-based services. Tg (time generation gap) for IN-based services whose introduction and replacement are mainly market driven will also decrease, but probably not to the same extent as introduction time, Ta. So, we can conclude that where services on the intelligent network platform are concerned, we may not have solved LSYD completely, but the intelligent network allows more dynamic introduction of, or change of, a service. This will, no doubt, make it easier to expand the number of services in future. Where the introduction and changing of intelligent network platforms is concerned, however, the same rules are in effect as before the intelligent network, as Ta is not changed (not decreased).

8.3 SOLUTIONS TO THE USER WILLINGNESS PROBLEM

Section 1.3 presented a hypothetical dialogue between a network operator and a customer. The conclusion was that even if the network operator provides a good user interface and good services, something is still missing. I called this user willingness (UW). I also said that UW could not be created unless user friendly terminals and interfaces were provided. But even this is not enough. How do we create UW?

In Section 1.3, we said that one reason for a lack of UW is peoples' habits. Special routines at work or at home are very difficult to change. Or, in other words, if people are currently using a nontelecommunications solution, having a network operator advertise and provide information about services is not enough. Old habits make it difficult for users to change to a telecommunications solution. The resistance will, however, be less from the younger generation.

Personally, I am convinced that we have needs in our daily lives that are best satisfied by a telecommunications solution. But, to be honest, I also believe that there are situations in which the opposite is true, that is, we have needs that are better solved outside the telecommunications network. However, and this is important, *there are more needs that can be better solved by telecommunications than the opposite.* Moreover, these needs will increase extremely rapidly in coming years. Examples, involving both private and business uses, include financial transactions, such as transferring funds between accounts; remote control, such as turning on the heat in your winter cottage before you arrive; and database searches.

Let us assume that good services, good availability, and acceptable prices are already provided by the telecommunications network. How do we then get people to switch to the telecommunications solutions that, in most cases, are better for them? I think we need to consider three key factors in attaining the goal of increasing UW, which are described in the following sections.

A more personal way of using the network. Let us return for a while to the time before the automation of exchanges, to the beginning of the century, when all communication was manually controlled by operators. Telecommunication at this time was mostly *local*, that is, it involved people within the same village or town. The operator was often situated in the middle of the village, and she (it was usually a woman) connected you to the called party you requested. Because of her physical situation in the middle of the village, she was often able to maintain personal contact with the inhabitants (the subscribers) and perhaps even see them. If someone wanted to speak to the tailor, for example, she could either connect the calling party to his home or to the barber shop, as she had seen the tailor step in for a haircut.

This is an example of something we lack today. In fact, despite high-technology systems with digitalization and stored program control exchanges and intelligent networks, we cannot match the service that the village operator could provide 70–90 years ago. If we could provide this, I do not think we would need to discuss matters like UW in the 1990s.

But, to be fair, as we discussed in Chapter 2, our method of communication has shifted *from local to global* communication without any geographical limitations, which drastically increases the complexity. And, although we do not wish to go back to a system of operators establishing connections manually, we must try to provide the benefits of the old system in today's system. Examples of this include calling by name or function ("The tailor, please"). A personal communication network (PCN) based on the intelligent network forms the base for a future solution to this. We cannot achieve a total solution that will satisfy everyone at once, but instead must proceed in a step-by-step evolution of PCN towards *universal personal telephony* (UPT).

The habits of the customers must be changed. Changing people's habits is one of the most difficult tasks we can undertake. This is also true if you are trying to get people to move from a nontelecommunications to a telecommunications solution. There are two main ways to get people to change a habit: you can *convince* them or you can *force* them. Of course, convincing a person is much better, but we must remember that many successful changes have been made by forcing individuals to change.

But, before discussing how to bring this change about, we must consider the fact that if we want people to use telecommunications services on a broad scale, a few basic things must be offered:

- Availability of services must be high,
- Service must be perceived as being better than the alternatives,
- The price must be competitive with the alternatives.

Take availability, for example. No one could accept the same low level of availability for telecommunications services that we have for cash-dispensing machines. We often fail to complete a transaction because the line to the computer system is down or the machine is out of money (especially on weekends).

But, despite the low level of availability, the introduction of cash-dispensing machines has been successful. People use them very frequently, instead of queuing up at the bank cashier. Why is this so? Banks have, through a combination of convincing and forcing, succeeded in changing the habits of the hurried people of the 80s and 90s in two ways:

1. By mounting an enormous advertising campaign that convinces customers they can
   - Save time by not queuing in front of a cashier in the bank,
   - Withdraw money outside of banking hours,
   - Get account balance receipts with withdrawn money.
2. By being somewhat "mean" to customers and forcing them to use cash dispensing machines by
   - Reducing the hours banks are open,
   - Charging customers for "counter" services the bank does not want to offer in the long run,
   - Providing all customers (new and old) with a card for cash dispensing machines.

Before long, cash dispensing machines are so widely used that they practically sell themselves.

When changing people's habits and trying to create UW for services in telecommunications, I believe that we must *prioritize* in order to convince and *provide information* instead of forcing. (by using marketing campaigns, TV spots, and so on).

A good example of the power of information is when people learn to make a *call forwarding unconditional* (CFU) on their home phone to their weekend cottage before going there. Normally, people do not bother to do this. However, once they have been informed about the benefits, such as being able to receive calls from friends that they would have missed before, the use of CFU increases. Many people also use this function to keep burglars away from their homes during the weekend. Burglars often call first to see if anyone is at home, and if you can answer your home telephone while at another location, no one knows that you are not at home.

These were two examples of how people can *realize the benefits* of using telecommunications services.

User interfaces should be harmonized across systems and be more functional. The lack of harmony in procedures that users on different systems employ to access the same services hinders UW. And, of course, we still face the problem of users on the public networks employing different procedures for systems from different network operators.

There is, however, a bigger problem, which is the lack of harmony of user procedures for services between public networks and private exchanges. At home, most of us are able to (or will soon be able to) use many value-added services on the public networks. But what happens when we are at work? There, perhaps 50% or more of us are connected to a PABX, which probably has a different set of services than the public network. Moreover, for services that exist on both systems, there are often different user procedures (different user codes) for using them.

Often, there is an additional problem encountered by PABX users when trying to reach the public network. Signaling limitations often make it impossible for them to use many of the services on the public network because the user codes cannot pass the PABX to the public network.

On many networks, centrex, wide area centrex (WAC), and virtual private networks (VPN) have been implemented. From the users' point of view, these will act in the same manner as private systems (like PABXs and private networks). This further underscores the need for harmony between private and public networks with respect to user procedures for services.

8.4 CONCLUSIONS

The conclusions about large telecommunications systems that can be drawn from Sections 8.1–8.3 are as follows:

- The IN-based method of creating services by combining service-independent building blocks (SIBs) on a stable platform partly solves the LSYD. This makes it possible to introduce a new service rapidly on the technical platform. But, as was stressed in several parts of the book, this must be complemented by the same fast introduction of the service in systems for administration and management. If we accomplish this, we can introduce a service much faster than before the intelligent network and drastically reduce the total introduction time of a new service, which is necessary for avoiding the LSYD.

- UW can be created by introducing a more personal way of using services that harmonizes them with users' daily lives. This also includes changes and harmonization in user interfaces and terminals, as well as changes in user habits.

Appendix

This appendix describes briefly service-independent building blocks (SIBs), service features (SFs), and supported services in capability set 1 (CS1) from ITU-T. (Autumn 1993). For detailed descriptions, please refer to the standards themselves.[1]

A.1 SERVICE-INDEPENDENT BUILDING BLOCKS DEFINED IN CAPABILITY SET 1

CS1 includes 13 SIBs (ITU-T recommendation Q.1213), which represent a minimum set required to define the CS1 targeted services in ITU-T recommendation Q.1211, described in Section A.3. The basic call process (BCP) has been defined as a specialized (14th) SIB. The CS1 from ETSI also includes, in addition to these 14 SIBs, another seven SIBs (for full descriptions, please refer to the standards).

SIB	Description
Algorithm	Applies a mathematical algorithm to data to produce a data result. May be used for reducing or increasing counters, for example, for mass calling services (televoting, and so on).
Basic call process (BCP)	A specialized SIB that provides basic call capabilities.
Charge	Allows customizing of charging rates for situations that require different rates than the basic call process. Examples are free of charge and fixed rate.
Compare	Compares two values, if greater than, equal to, or less than. Used for checking a current time against time and date routing, comparing a current value of number of calls with a predetermined value, and so on.

---

1. Full text of the source material may be obtained from the ITU Sales Section, Place de Nations, CH-1211 Geneva 20, Switzerland, Tel. +41 22 730 51 11, Fax +41 22 730 5194.

SIB	*Description*
Distribution	Distributes calls to different destinations according to a percentage distribution or time and date routing.
Limit	Limits the number of calls for a service.
Log call information	Logs detailed information about a call in a separate file. Information can be the time of the call, the time the call ended, the dialed number, the destination number, the calling line identification, the queuing time for the call, and so on.
Queue	Provides sequencing of calls to be completed to a called party. Provides the user with the capability to pass a call if resources are available, queue a call if resources are unavailable, play a recorded message to the caller during the queuing time, release the call from the queue when resources become available. The maximum number of calls that can be queued or the maximum time calls can be queued may be specified.
Screen	Checks to ensure that a number is in a specified list, for example, checks PIN code, account number, destination number. Used for functions like call screening and account calling (with or without card).
Service data management	Fetches, replaces, increases, or decreases end user–specific da
Status notification	Checks the status of a resource, to wait for the right status, to supervise and save changes in status, or to end the supervision of a resource.
Translate	Translates input information into output information, for example, translates a number into one or more numbers. Used for number translation services like freephone and for abbreviated dialing, call forwarding, and call transfer services.
User interaction	Allows information to be exchanged between the network and involved parties (calling or called parties). Used for playing an announcement machine recording and for receiving digits from the parties involved. Can also be used to decide the number of times a recorded message should be played. The message may be interrupted when digits are dialed.

Verify   Compare collected information with expected format. Often used after the SIB: user interaction.

A.2 CAPABILITY SET 1 TARGET SET OF SERVICE FEATURES

Table A.1 (first and second part) shows the mapping between services and service features in CS1 (ITU-T recommendation Q.1211).

Table A.1
Q.1211 Mapping Between Services and Service Features

Services	Service Features (first part)																		
	ABD	ATTC	AUTZ	AUT	ACB	CD	CF	CFC	GAP	CHA	LIM	LOG	QUE	TRA	CW	CUG	COC	CPM	CRA
ABD	C											O						O	
ACC	C		C									O							
AAB	O		C									O							
CD						C						O						O	
CF							C					O						O	
CRD								O			O	O	O					O	O
CCBS					C							O			O				
CON												O				O	O		
CCC	O		C									O							
DCR						C						O						O	
FMD												O						O	
FPH			O			O		O	O		O	O	O					O	O
MCI												C							
MAS						O			O		O	O	O					O	O
OCS												O						O	
PRM						O		O	O		O	O	O					O	O
SEC				C								O						O	
SCF								C				O						O	
SPL				O				O	O		O	O	O					O	O
VOT				O					O		O	O	O					O	O
TCS												O						O	
UAN				O				O	O		O	O	O					O	O
UPT			C									O						O	O
UDR												O						O	
VPN	O	O	O	O		O				O		O	O	O		O	O	O	O

Table A.1 continued

Services	CRG	DUP	FMD	MAS	MMC	MWC	OFA	ONC	ONE	ODR	OCS	OUP	PN	PRMC	PNP	REVC	SPLC	TCS	TDR
								Service Features (second part)											
ABD		O																	
ACC											C								
AAB											C								
CD									C	O									O
CF																			
CRD									C										
CCBS																			
CON					O	C													
CCC												C							
DCR										O									O
FMD			C																
FPH	O	O		O					C	O	O	O				C			O
MCI											C								
MAS				C						O	O	O							O
OCS											C								
PRM	O	O							C	O	O	O		C					O
SEC																			
SCF																			
SPL	O	O							C	O	O						C		
VOT			C							O	O	O							O
TCS																		C	
UAN	O								C	O	O	O							O
UPT		O	C									O	C				C		O
UDR										O									O
VPN	O		O				O	O				O				C			O

C = Core   The particular service feature is fundamental to the service, i.e., in the absence of the service feature the name of the service does not make sense as a commercial offering to the service subscriber.

O = Optional   The service feature is not core, i.e., without this service feature the name of the service would still make sense as a commercial offering to the service subscriber. Therefore, the service feature can be regarded as an optional enhancement to the service.

Service Features Fully Supported in CS1

The following is a list of the service features that are fully supported in CS1 (for full descriptions, please refer to the standards).

Service Feature	*Description*
Abbreviated dialing (ABD)	Allows the definition of abbreviated dialing.
Attendant (ATTC)	Allows virtual private network (VPN) users to access an attendant position within the VPN.
Authentication (AUT)	Verifies that a user is allowed to exercise certain options.
Authorization code (AUTZ)	Allows a VPN user to override calling restrictions.
Call distribution (CD)	Specifies the percentage of calls to be distributed among two or more destinations.
Call forwarding (CF)	Forwards incoming calls to another number.
Call forwarding on busy/don't answer (CFC)	Forwards particular incoming calls if the called user is busy or does not answer within a specified number of rings.
Call gapping (GAP)	Restricts the number of calls that can be routed to a particular destination.
Call limiter (LIM)	Specifies a maximum number of calls that can be simultaneously routed to a particular destination.
Call logging (LOG)	Creates a log record for each call received at a specified number.
Call queuing (QUE)	Queues calls meeting busy or no answer within a predetermined time. An announcement is given to the calling party.
Closed user group (CUG)	Allows a user to be a member of a set of VPN users who are normally authorized to make and receive calls only within the group.
Customer profile management (CPM)	Allows users to manage their customer profiles in real time.
Customized recorded announcement (CRA)	Allows a call to be completed to a (customized) terminating announcement instead of a subscriber line.
Customized ringing (CRG)	Allocates a distinctive ringing to a list of calling parties.
Destinating user prompting (DUP)	Prompts called parties with a specific announcement.
Follow-me diversion (FMD)	Routes incoming calls to a new location.

Service Feature	Description
Mass calling (MAS)	Allows the processing of huge numbers of incoming calls, for example, calls generated by broadcasted advertising or games.
Off net access (OFA)	Allows VPN users to access the VPN from any non-VPN exchange.
Off net calling (ONC)	Allows VPN users to call outside the VPN.
One number (ONE)	Permits two or more terminating lines, also at different locations, to have a single telephone number.
Origin-dependent routing (ODR)	Enables users to accept or reject calls and, if they acceptance a call, route the call according to the calling party's geographical location.
Originating call screening (OCS)	Allows users to bar calls from certain areas.
Originating user prompter (OUP)	Allows users to provide an announcement requesting calling parties to enter a digit or series of digits.
Personal numbering (PN)	Supports a universal personal telephony (UPT) number that uniquely identifies each UPT user and is used by calling parties to reach the UPT user.
Premium charging (PRMC)	Allows for the payback of part of the cost of a call to a value-added service provider.
Private numbering plan (PNP)	Allows subscribers to maintain a numbering plan within their private network that is separate from the public network.
Reverse charging (REVC)	Allows subscribers to be charged for the entire cost of calls they receive.
Split charging (SPLC)	Allows calling and called parties to share the cost of a call.
Terminating call screening (TCS)	Screen calls based on the terminating number called.
Time-dependent routing (TDR)	Enables users to accept or reject calls and, if they accept a call, route the call according to time and date. (Other treatments than routing can be done, according to time and dates.)

Service Features Partially Supported in CS1

The following is a list of the service features that are partially supported in CS1 (for full descriptions, please refer to the standards).

Service Feature	Description
Automatic call back (ACB)	Allows the called party to automatically call back the calling party of the last call.
Call hold with announcement (CHA)	Allows subscribers to place a call on hold with options to play music or a customized message.
Call transfer (TRA)	Allows a user to place a call on hold and transfer the call to another location.
Call waiting (CW)	Notifies the called party, when on a call, that another party is trying to reach him or her.
Consultation calling (COC)	Allows the user to place a call on hold, in order to initiate a new call for consultation.
Meet-me conference (MMC)	Reserves a conference resource for making a multiparty call, indicating the date, time, and duration of the conference.
Multiway calling (MWC)	Establishes multiple, simultaneous calls among parties.

A.3 CAPABILITY SET 1 TARGET SET OF SERVICES

These services should be considered as a minimum set of services for CS1 (ITU-T recommendation Q.1211). Table A.1 (first and second part) shows the mapping between services and service features in CS1. (For full descriptions, please refer to the standards.)

Services Fully Supported in CS1

The following services are fully supported in CS1.

Service	Description
Abbreviated dialing (ABD)	Allows abbreviated versions of telephone numbers. Translation to the real number is done by the network or by the terminal.
Account card calling (ACC)	Allows users to make calls from any terminal (with or without card reading facilities) and have the call charged to an account rather than the terminal used. Used with a secret PIN code.
Automatic alternative billing (AAB)	Charges calls made from any terminal to an account rather than to the calling or called party.
Call distribution (CD)	Routes incoming calls to different destinations based on specified criteria.
Call forwarding (CF)	Forwards calls to another number.

Service	Description
Call rerouting distribution (CRD)	Allows users to have incoming calls encounter a triggering condition.
Credit card calling (CCC)	Charges calls made from any access point to a specified account number.
Destination call routing (DCR)	Routes calls to destinations according to specified criteria, such as time, date, originating area, or calling line identity.
Follow-me diversion (FMD)	Allows users to remotely control redirection of incoming calls to other locations.
Freephone (FPH)	Allows reverse charging.
Malicious call identification (MCI)	Logs incoming calls.
Mass calling (MAS)	Allows the network operator to temporarily allocate a single number to a user. Includes instantaneous high-volume traffic routed to one or multiple locations (for example, televoting).
Originating call screening (OCS)	Specifies that outgoing calls may be restricted or allowed.
Premium rate (PRM)	Allows for the payback of part of the cost of a call to a value-added service provider.
Security screening (SEC)	Screens network users attempting to access the subscriber's network, systems, or applications.
Selective call forwarding on busy/no answer (SCF)	Allows called parties to forward preselected calls if busy or don't answer.
Split charging (SPL)	Allows calling and called parties to split call charges.
Televoting (VOT)	Allows users to vote by phone. Calling parties call one of two or more numbers to cast their vote.
Terminating call screening (TCS)	Specifies that incoming calls be either restricted or allowed.
Universal access number (UAN)	Allows a user with several terminating lines in any number of locations to be reached by one number.
Universal personal telecommunications (UPT)	Allows users to access telecommunication services via unique personal telecommunication number (PTN) across multiple networks at any network access.
User-defined routing (UDR)	Allows users to specify how outgoing calls shall be routed.
Virtual private network (VPN)	Allows users to build a private network with public network resources.

Services Partially Supported in CS1

The following two services are only *partly supported* in CS1.

Service	*Description*
Call completion to busy subscriber (CCBS)	Allows calling parties who encounter a busy destination to be informed when the busy destination becomes free, without having to make a new call attempt.
Conference calling (CON)	Connects multiple parties in a single conversation.

About the Author

Jan Thörner has worked at Telia (formerly Swedish TeleCom) in Stockholm since 1989, where he is responsible for coordinating research activities for services and related areas, including intelligent networks, in narrowband networks (PSTN and narrowband ISDN). He is also the Swedish project leader in the intelligent network cooperation between the Netherlands, Switzerland, and Sweden. Thörner holds a Master of Science degree from the Royal Institute of Technology in Stockholm. He has broad experience in system and software design for both telecommunications and data communications and has been employed in these areas by Ericsson, Philips, and Telia. In particular, he has focused on performance, response times, and optimization problems in public telephone networks and local area networks. He was responsible for introducing the first two SCPs on the Swedish network in 1990.

Index

The Artech House Telecommunications Library

Vinton G. Cerf, Series Editor

For further information on these and other Artech House titles, contact:

Artech House
685 Canton Street
Norwood, MA 02062
781-769-9750
Fax: 781-769-6334
Telex: 951-659
email: artech@artech-house.com

Artech House
Portland House, Stag Place
London SW1E 5XA England
+44 (0) 171-973-8077
Fax: +44 (0) 171-630-0166
Telex: 951-659
email: artech-uk@artech-house.com

WWW: http://www.artech-house.com